Official
TELECOMMUNICATIONS
Dictionary

Legal and Regulatory Definitions

Thomas F. P. Sullivan, Editor

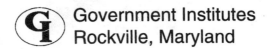

Government Institutes
Rockville, Maryland

Government Institutes, Inc., 4 Research Place, Rockville, Maryland 20850
Phone: (301) 921-2355
Fax: (301) 921-0373
Email: giinfo@govinst.com
Internet address: http://www.govinst.com

ISBN: 0-86587-564-2

Printed in the United States of America

Preface

This dictionary contains the *official* legal and regulatory definitions from Congressional telecommunication legislation and Agency regulations. The terms and acronyms were taken from the following sources:

(1) Telecommunications Act of 1996
Public Law 104-104, signed February 8, 1996
(47 USC 151 et seq.)

(2) Communications Act of 1934
as amended by the Telecommunications Act of 1996
(47 USC 151 et seq.)

(3) Title 47 of the Code of Federal Regulations
as issued by the Federal Communications Commission.

All of these official terms and acronyms carry citations to the above sources.

Some terms have multiple definitions. If the definitions are different, they are listed with a bracketed number after the term, e.g., [1], [2], etc. The terms with multiple definitions are listed in order of occurrence in the United States Code (USC) and Code of Federal Regulations (CFR).

If the terms are the same but in multiple cites, the cites are listed in USC then CFR order at the end of the definition.

Some terms carry additional clarification of the definition sources such as:

(1) CONV - International Telecommunications Convention, Malaga-Torremolinos, 1973.

(2) RR - Radio Regulations, Geneva, 1982.

(3) FCC - Federal Communications Commission

Acronyms are in alphabetical order throughout the dictionary and are followed by one cite. This cite notes only the first occurrence of the acronym in the USC or CFR.

Official definitions of technical terms used in telecommunication standards can be found in an additional publication from Government Institutes, *Telecommunications: Glossary of Telecommunication Terms* (FED-STD-1037C).

We welcome suggestions on future editions and sincerely hope that this will provide a useful tool.

Thomas F.P. Sullivan
President, Government Institutes

A

A/A1

Connections to the "hold functions of key telephone systems which use such connections. In such systems, the "A" lead corresponding to a particular telephone line is shorted to the "A1" lead when that line is placed in the "off-hook" state to permit proper operation of the "hold" functions associated with that line. [47 CFR 68.502]

AAIC

Accounting authority identification codes [47 CFR 80.95]

AAT

Above average terrain [47 CFR 90.307]

AATE

Assumed average terrain elevation [47 CFR 90.279]

ABC

American Broadcasting Company [47 CFR 11.43]

AC

Alternating current [47 CFR 15.107]

accelerated depreciation

A depreciation method or period of time, including the treatment given cost of removal and gross salvage, used in calculating depreciation deductions on income tax returns which is different from the depreciation method or period of time prescribed by the Commission for use in calculating depreciation expense recorded in a company's books of account. [47 CFR 32.9000]

accepted interference

Interference at a higher level than defined as permissible interference and which has been agreed upon between two or more administrations without prejudice to other administrations. (RR) [47 CFR 2.1]

access code

A sequence of numbers that, when dialed, connect the caller to the provider of operator services associated with that sequence. [47 USC 226, 47 CFR 64.708]

access line

A communications facility extending from a customer's premises to a serving central office comprising a subscriber line and, if necessary, a trunk facility, *e.g*, a WATS access line, TWX access line. [47 CFR Pt. 36, App.]

access minutes (access minutes of use)

That usage of exchange facilities in interstate or foreign service for the purpose of calculating chargeable usage. On the originating end of an interstate or foreign call, usage is to be measured from the time the originating end user's call is delivered by the telephone company and acknowledged as received by the interexchange carrier's facilities connected with the originating exchange. On the terminating end of an interstate or foreign call, usage is to be measured from the time the call is received by the end user in the terminating exchange. Timing of usage at both the originating and terminating end of an interstate or foreign call shall terminate when the calling or called party disconnects, whichever event is recognized first in the originating and terminating end exchanges, as applicable. [47 CFR 69.2]

access service

Services and facilities provided for the origination or termination of any interstate or foreign telecommunication. [47 CFR 69.2]

access software

Software (including client or server software) or enabling tools that do not create or provide the content of the communication but that allow a user to do any one or more of the following: (A) filter, screen, allow, or disallow content; (B) pick, choose, analyze, or digest content; or (C) transmit, receive, display, forward, cache, search, subset, organize, reorganize, or translate content. [47 USC 223]

access software provider

A provider of software (including client or server software), or enabling tools that do any one or more of the following: (A) filter, screen, allow, or disallow content: (B) pick, choose, analyze, or digest content; or (C) transmit, receive, display, forward, cache, search, subset, organize, reorganize, or translate content. [47 USC 230]

account

A specific element of a chart of accounts used to record, classify and accumulate similar financial transactions resulting from the operations of the entity. "Accounts" or "these accounts" refer to the accounts of this system of accounts. [47 CFR 32.9000]

account xxxx

The account of that number kept in accordance with the uniform system of accounts for Class A and Class B Telecommunications Companies in 47 CFR part 32. [47 CFR 65.810]

accounting system

The total set of interrelated principles, rules, requirements, definitions, accounts, records, procedures and mechanisms necessary to operate and evaluate the entity from a financial perspective. An accounting system generally consists of a chart of accounts, various parallel subsystems and subsidiary records. An accounting system is utilized to provide the necessary financial information to users to meet judiciary and other responsibilities. [47 CFR 32.9000]

accredited standards development organization

An entity composed of industry members which has been accredited by an institution vested with the responsibility for standards accreditation by the industry. [47 USC 273]

Act

The Communications Act of 1934 (48 Stat. 1004; 47 U.S.C. chapter 5), as amended. [47 CFR 61.3]

activated channels

Those channels engineered at the headend of a cable system for the provision of services generally available to residential subscribers of the cable system, regardless of whether such services actually are provided, including any channel designated for public, educational or governmental use. [47 USC 522, 47 CFR 76.5]

active satellite [1]

A satellite carrying a station intended to transmit or retransmit radio communication signals. (RR) [47 CFR 2.1]

active satellite [2]

An earth satellite carrying a station intended to transmit or retransmit radiocommunication signals. [47 CFR 25.201]

active sensor

A measuring instrument in the earth exploration-satellite service or in the space research service by means of which information is obtained by transmission and reception of radio waves. (RR) [47 CFR 2.1]

actual price index (API)

An index of the level of aggregate rate element rates in a basket, which index is calculated pursuant to §61.46. [47 CFR 61.3]

ADI

Area of dominant influence [47 CFR 73.525, 76.56]

adjudicative proceeding

Any proceeding, other than a rule making or a tariff proceeding involving future rates or practices, initiated upon the Commission's own motion or upon the filing of an application, a petition for special relief or waiver, or a complaint or similar pleading that involves the determination of rights and responsibilities of specific parties. [47 CFR 1.202]

administration

Any governmental department or service responsible for discharging the obligations undertaken in the Convention of the International Telecommunication Union and the Regulations. (CONV) [47 CFR 2.1]

administrative offset

Withholding money payable by the United States Government to, or held by the government for, a person to satisfy a debt the person owes the government. [47 CFR 1.1901]

ADR

Administrative message [47 CFR 11.31]

advanced television services

Television services provided using digital or other advanced technology as further defined in the opinion, report, and order of the Commission entitled Advanced Television Systems and Their Impact Upon the Existing Television Broadcast Service, MM Docket 87-268, adopted September 17, 1992, and successor proceedings. [47 USC 335]

adverse finding and adverse final action

As used in paragraph (c) of this section include adjudications made by an ultimate trier of fact, whether a government agency or court, but do not include factual determinations which are subject to review *de novo* unless the time for taking such review has expired under the relevant procedural rules. The pendency of an appeal of an adverse finding or adverse final action does not relieve a permittee or licensee from its obligation to report the finding or action. [47 CFR 1.65]

advertisement

Any message or other programming material which is broadcast or otherwise transmitted in exchange for any remuneration, and which is intended (1) to promote any service, facility, or product offered by any person who is engaged in such offering for profit; (2) to express the views of any person with respect to any matter of public importance or interest; or (3) to support or oppose any candidate for political office. [47 USC 399b]

aeronautical advisory station (unicom)

An aeronautical station used for advisory and civil defense communications primarily with private aircraft stations. [47 CFR 87.5]

aeronautical earth station

An earth station in the fixed-satellite service, or, in some cases, in the aeronautical mobile-satellite service, or, in some cases, in the aeronautical mobile-satellite service, located at a specified fixed point on land to provide a feeder link for the aeronautical mobile-satellite service. (RR) [47 CFR 2.1]

aeronautical en route station

An aeronautical station which communicates with aircraft stations in flight status or with other aeronautical en route stations. [47 CFR 87.5]

aeronautical fixed service

A radiocommunication service between specified fixed points provided primarily for the safety of air navigation and for the regular, efficient and economical operation of air transport. A station in this service is an aeronautical fixed station. [47 CFR 2.1, 87.5]

aeronautical fixed station

A station in the aeronautical fixed service. (RR) [47 CFR 2.1]

aeronautical mobile off-route (OR) service

An aeronautical mobile service intended for communications, including those relating to flight coordination, primarily outside national or international civil air routes. (RR) [47 CFR 2.1, 87.5]

aeronautical mobile route (R) service

An aeronautical mobile service reserved for communications relating to safety and regularity of flight, primarily along national or international civil air routes. (RR) [47 CFR 2.1, 87.5]

aeronautical mobile-satellite off-route (OR) service

An aeronautical mobile-satellite service intended for communications, including those relating to flight coordination, primarily outside national and international civil air routes. (RR) [47 CFR 2.1, 87.5]

aeronautical mobile-satellite route (R) service.

An aeronautical mobile-satellite service reserved for communications relating to safety and regularity of flights, primarily along national or international civil air routes. (RR) [47 CFR 2.1]

aeronautical mobile-satellite service [1]

A mobile-satellite service in which mobile earth stations are loaded on board aircraft; survival craft stations and emergency position-indicating radio-beacon stations may also participate in this service. (RR) [47 CFR 2.1]

aeronautical mobile-satellite service [2]

A mobile-satellite service in which mobile earth stations are located on board aircraft. [47 CFR 87.5]

aeronautical mobile service

A mobile service between aeronautical stations and aircraft stations, or between aircraft stations, in which survival craft stations may participate; emergency position-indicating radio beacon stations may also participate in this service on designated distress and emergency frequencies (RR) [47 CFR 2.1, 87.5]

aeronautical multicom station

An aeronautical station used to provide communications to conduct the activities being performed by, or directed from, private aircraft. [47 CFR 87.5]

aeronautical radionavigation-satellite service

A radio-navigation-satellite service in which earth stations are located on board aircraft (RR) [47 CFR 2.1]

aeronautical radionavigation service

A radio-navigation service intended for the benefit and for the safe operation of aircraft (RR) [47 CFR 2.1, 87.5]

aeronautical search and rescue station

An aeronautical station for communication with aircraft and other aeronautical search and rescue stations pertaining to search and rescue activities with aircraft. [47 CFR 87.5]

aeronautical station

A land station in the aeronautical mobile service [47 CFR 2.1]. In certain instances an aeronautical station may be located, for example, on board ship or on a platform at sea. [47 CFR 87.5]

aeronautical utility mobile station

A mobile station used on airports for communications relating to vehicular ground traffic. [47 CFR 87.5]

affiliate [1]

A person that (directly or indirectly) owns or controls, is owned or controlled by, or is under common ownership or control with, another person. For purposes of this paragraph, the term "own means to own an equity interest (or the equivalent thereof) of more than 10 percent. [47 USC 153]

affiliate [2]

Shall have the same meaning as in Section 3 of this Act, except that, for purposes of paragraph (1)(B)(i) an aggregate voting equity interest in Bell Communications Research, Inc, of at least 5 percent of its total voting equity, owned directly or indirectly by more than 1 otherwise unaffiliated Bell operating company, shall constitute an affiliate relationship; and (ii) a voting equity interest in Bell Communications Research, Inc., by any otherwise unaffiliated Bell operating company of less than 1 percent of Bell Communications Research's total voting equity shall not be considered to be an equity interest under this paragraph. [47 USC 273]

affiliate [3]

Any entity that, directly or indirectly, owns or controls, is owned or controlled by, or is under common ownership or control with, a Bell operating company. Such term shall not include a separated affiliate. [47 USC 274]

affiliate [4]

When used in relation to any person, another person who owns or controls, is owned or controlled by, or is under common ownership or control with, such person. [47 USC 522, 47 CFR 76.5]

affiliate [5]

Determinations regarding whether an individual or entity will be considered an affiliate of (1) an applicant or (2) a person holding an attributable interest in an applicant under paragraph (b)(2) will be made pursuant to the general affiliation rules set form in §24.710(1). [47 CFR 24.230]

affiliate [6]

(1) Basis for Affiliation. An individual or entity is an affiliate of an applicant or of a person holding an attributable interest in an applicant (both referred to herein as "the applicant") if such individual or entity (i) directly or indirectly controls or has the power to control the applicant, or (ii) is directly or indirectly controlled by the applicant, or (iii) is directly or indirectly controlled by a third party or parties that also controls or has the power to control the applicant, or (iv) has an "identity of interest" with the applicant. [47 CFR 24.720];

(2) Nature of Control in Determining Affiliation. (i) Every business concern is considered to have one or more parties who directly or indirectly control or have the power to control it. Control may be affirmative or negative and it is immaterial whether it is exercised so long as the power to control exists; (ii) control can arise through stock ownership; occupancy of director, officer or key employee positions; contractual or other business relations; or combinations of these and other factors. A key employee is an employee who, because of his/her position in the concern, has a critical influence in or substantive control over the operations or management of the concern; (iii) control can arise through management positions where a concern's voting stock is so widely distributed that no effective control can be established.

(3) Identity of interest between and among persons. Affiliation can arise between or among two or more persons with an identity of interest, such as members of the same family or persons with common investments. In determining if the applicant controls or is controlled by a concern, persons with an identity of interest will be treated as though they were one person. [47 CFR 90.814]

affiliated

For purposes of determining whether a video programming vendor is "affiliated" with a multichannel video programming distributor, as used in this subpart, the definitions for "attributable interest" contained in the notes to §76.501 shall be used provided, however that (1) the single majority shareholder provisions of Note 2(b) to §76.501 and the limited partner insulation provisions of Note 2(g) to §76.501 shall not apply; and (2) the provisions of Note 2(a) to §76.501 regarding five (5) percent interests shall include all voting or nonvoting stock or limited partnership equity interests of five (5) percent or more. [47 CFR 76.1300]

affiliated companies

Companies that directly or indirectly through one or more intermediaries, control or are controlled by, or are under common control with, the accounting company. See also "Control." [47 CFR 32.9000]

affiliation arising under stock options, convertible debentures, and agreements to merge

Stock options, convertible debentures, and agreements to merge (including agreements in principle) are generally considered to have a present effect on the power to control the concern. Therefore, in making a size determination, such options, debentures, and agreements will generally be treated as though the rights held thereunder had been exercised. However, neither an affiliate nor an applicant can use such options and debentures to appear to terminate its control over another concern before it actually does so. [47 CFR 24.720, 90.814]

affiliation through common facilities

Affiliation generally arises where one concern shares office space and/or employees and/or other facilities with another concern, particularly where such concerns are in the same or related industry or field of operations, or where such concerns were formerly affiliated, and through these sharing arrangements one concern has control, or potential control, of the other concern. [47 CFR 24.720, 90.814]

affiliation through common management

Affiliation generally arises where officers, directors, or key employees serve as the majority or otherwise as the controlling element of the board of directors and/or the management of another entity. [47 CFR 24.720, 90.814]

affiliation through contractual relationships

Affiliation generally arises where one concern is dependent upon another concern for contracts and business to such a degree that one concern has control, or potential control, of the other concern. [47 CFR 24.720, 90.814]

affiliation through stock ownership

(i) An applicant is presumed to control or have the power to control a concern if he or she owns or controls or has the power to control 50 percent or more of its voting stock.

(ii) An applicant is presumed to control or have the power to control a concern even through he or she owns, controls or has the power to control less than 50 percent of the concern's voting stock, if the block of stock he or she owns, controls or has the power to control is large as compared with any other outstanding block of stock.

(iii) If two or more persons each owns, controls or has the power to control less than 50 percent of the voting stock of a concern, such minority holdings are equal or approximately equal in size, and the aggregate of these minority holdings is large as compared with any other stock holding, the presumption arises that each one of these persons individually controls or has the power to control the concern; however, such presumption may be rebutted by a showing that such control or power to control, in fact, does not exist. [47 CFR 24.720, 90.814]

affiliation under joint venture arrangements

(i) A joint venture for size determination purposes is an association of concerns and/or individuals, with interests in any degree or proportion, formed by contract, express or implied, to engage in and carry out a single, specific business venture for joint profit for which purpose they combine their efforts, property, money, skill and knowledge, but not on a continuing or permanent basis for conducting business generally. The determination whether an entity is a joint venture is based upon the facts of the business operation, regardless of how the business operation may be designated by the parties involved. An agreement to share profits/losses proportionate to each party's contribution to the business operation is a significant factor in determining whether the business operation is a joint venture.

(ii) The parties to a joint venture are considered to be affiliated with each other. [47 CFR 24.720, 90.814]

affiliation under voting trusts

(i) Stock interests held in trust shall be deemed controlled by any person who holds or shares the power to vote such stock, to any person who has the sole power to sell such stock, and to any person who has the right to revoke the trust at will or to replace the trustee at will.

(ii) If a trustee has a familial, personal or extra-trust business relationship to the grantor or the beneficiary, the stock interests held in trust will be deemed controlled by the grantor or beneficiary, as appropriate.

(iii) If the primary purpose of a voting trust, or similar agreement, is to separate voting power from beneficial ownership of voting stock for the purpose of shifting control of or the power to control a concern in order that such concern or another concern may meet the Commission's size standards, such voting trust shall not be considered valid for this purpose regardless of whether it is or is not recognized within the appropriate jurisdiction. [47 CFR 24.720, 90.814]

AFRTS

Armed forces radio and television services [47 CFR 73.1207]

AFUDC

Allowance for funds used during construction [47 CFR 32.2000]

Agency [1]

(1) The Commission, (2) a board of Commissioners (see §0.212), (3) the Telecommunications Committee (see §0.215), and (4) any other group of Commissioners hereafter established by the Commission on a continuing or *ad hoc* basis and authorized to act on behalf of the Commission. [47 CFR 0.601]

Agency [2]

The Federal Communications Commission (Commission) or any other agency of the U.S. Government as defined by section 105 of title 5 U.S.C., the U.S. Postal Service, the U.S. Postal Rate Commission, a military department as defined by section 102 of title 5 U.S.C., an agency or court of the judicial branch, or and an agency of the legislative branch, including the U.S. Senate and the U.S. House of Representatives. [47 CFR 1.1901]

Agency [3]

The Federal Communications Commission. [47 CFR 1.3001]

agency ethics official, designated agency ethics official, employee, market value, person, and prohibited source

Have the same meaning as found in 5 CFR 2635.102, 2635.203. [47 CFR 1.3001]

Agency Head

The Chairman of the Federal Communications Commission. [47 CFR 1.1901]

agent [1]

     With respect to any person, includes any employee of such person. [47 USC 605]

agent [2]

     Includes furloughed, pensioned, and superannuated agents. [47 CFR 41.1]

AGRAS

     Air-ground radiotelephone automated service [47 CFR 22.819]

aggregate customer information

     Collective data that relates to a group or category of services or customers, from which individual customer identities and characteristics have been removed. [47 USC 222]

aggregate information

     Collective data that relate to a group or category of services or customers, from which individual customer identities or characteristics have been removed. [47 CFR 64.1600]

aggregator

     Any person that, in the ordinary course of its operations, makes telephones available to the public or to transient users of its premises, for interstate telephone calls using a provider of operator services. [47 USC 226, 47 CFR 64.708]

AIOD

     Automatic identified outward dialing [47 CFR 68.222]

AIOD data channel simulator

     A test circuit that simulates a telephone line during the idle and data-receiver-attached conditions of central office AIOD circuits. The schematic of Figure 68.3(g) is illustrative of the type of circuit that will be required; alternative implementations may be used provided that the same DC voltage and current characteristics and AC impedance characteristics will be presented to the AIOD equipment under test. When used, the simulator circuit shall be operated over the entire range of resistance, polarities and voltage limits indicated in Figure 68.3(g). Whenever DC current is changed, sufficient time shall be allocated for the current to reach a steady-state condition before continuing the test. [47 CFR 68.3]

AIOD leads

Terminal equipment leads at the interface solely to transmit Automatic Identified Outward Dialing (AIOD) data from a PBX to the public switched telephone network or to switched telephone network or to switched service networks (*e.g.*, EPSCS) so that a telephone company can provide a PBX customer with a detailed monthly bill identifying long distance usage by individual PBX stations, tie trunks or the attendant. Data on the channel is transmitted in only one direction, from the PBX to the central office, and consists of a trunk number and a station number for each outgoing call. Two-way dc simplex signaling, as defined for the terminal equipment by the data channel simulator circuit, is used to coordinate the transmitting and receiving functions. One or more pairs of AIOD leads, each designated T (AI) and R (AI) to distinguish them from other tip and ring leads, may appear at an interface, depending on the number of central offices that process AIOD calls for the PBX. However, unless otherwise stated, these leads at the interface should be treated as telephone connections as defined in (x) of this section or as a tip and ring where the term "telephone connection" is not used. [47 CFR 68.3]

airborne station

A mobile station in the Air-Ground Radiotelephone Service authorized for use on aircraft while in flight or on the ground. [47 CFR 22.99]

air carrier aircraft station

A mobile station on board an aircraft which is engaged in, or essential to, the transportation of passengers or cargo for hire. [47 CFR 87.5]

aircraft earth station (AES)

A mobile earth station in the aeronautical mobile-satellite service located on board an aircraft. [47 CFR 2.1, 87.5]

aircraft station [1]

Has the meaning given it by the Commission by rule. [47 USC 403]

aircraft station [2]

A mobile station in the aeronautical mobile service, other than a survival craft station, located on board an aircraft. (RR) [47 CFR 2.1, 87.5]

air-ground radiotelephone service

A radio service in which common carriers are authorized to offer and provide radio telecommunications service for hire to subscribers in aircraft. [47 CFR 22.99]

airport

An area of land or water that is used or intended to be used for the landing and takeoff of aircraft, and includes its buildings and facilities, if any. [47 CFR 87.5]

airport control tower (control tower) station

An aeronautical station providing communication between a control tower and aircraft. [47 CFR 87.5]

alarm monitoring service

A service that uses a device located at a residence, place of business, or other fixed premises (1) to receive signals from other devices located at or about such premises regarding a possible threat at such premises to life, safety, or property, from burglary, fire, vandalism, bodily injury or other emergency, and (2) to transmit a signal regarding such threat by means of transmission facilities of a local exchange carrier or one of its affiliates to a remote monitoring center to alert a person at such center of the need to inform the customer or another person or police, fire, rescue, security, or public safety personnel of such threat, but does not include a service that uses a medical monitoring device attached to an individual for the automatic surveillance of an ongoing medical condition. [47 USC 275]

Alaska—private fixed station

A fixed station in Alaska which is licensed by the Commission for radio communication within Alaska and with associated ship stations, on single frequency channels. Alaska-private fixed stations are also eligible to communicate with Alaska-public fixed stations on paired channels. [47 CFR 80.5]

Alaska—public fixed station

A fixed station in Alaska which is open to public correspondence and is licensed by the Commission for radio communication with Alaska-private fixed stations on paired channels. [47 CFR 80.5]

allocation (of a frequency band)

Entry in the Table of Frequency Allocations of a given frequency band for the purpose of its use by one or more terrestrial or space radiocommunication services or the radio astronomy service under specified conditions. This term shall also be applied to the frequency band concerned. (RR) [47 CFR 2.1]

allotment (of a radio frequency or radio frequency channel)

Entry of a designated frequency channel in an agreed plan, adopted by a competent conference, for use by one or more administrations for a terrestrial or space radiocommunication service in one or more identified countries or geographical area and under specified conditions. (RR) [47 CFR 2.1]

altitude of apogee or perigee

The altitude of the apogee or perigee above a specified reference surface serving to represent the surface of the earth. (RR) [47 CFR 2.1]

AM

Amplitude modulation [47 CFR 73.681]

AM broadcast band

The band of frequencies extending from 535 to.1705 kHz. [47 CFR 73.14]

AM broadcast channel

The band of frequencies occupied by the carrier and the upper and lower sidebands of an AM broadcast signal with the carrier frequency at the center. Channels are designated by their assigned carrier frequencies. The 117 carrier frequencies assigned to AM broadcast stations begin at 540 kHz and progress in 10 kHz steps to 1700 kHz. (See §73.21 for the classification of AM broadcast channels). [47 CFR 73.14]

AM broadcast station
     A broadcast station licensed for the dissemination of radio communications intended to be received by the public and operated on a channel in the AM broadcast band. [47 CFR 73.14]

amateur operator
     A person holding a written authorization to be the control operator of an amateur station.   [47 CFR 97.3]

amateur radio services
     The amateur service, the amateur-satellite service and the radio amateur civil emergency service. [47 CFR 97.3]

amateur-satellite service
     A radiocommunication service using space stations on earth satellites for the same purposes as those of the amateur service. (RR) [47 CFR 2.1, 97.3]

amateur service
     A radiocommunication service for the purpose of self-training, intercommunication and technical investigations carried out by amateurs, that is, by duly authorized persons interested in radio technique solely with a personal aim and without pecuniary interest. (RR) [47 CFR 2.1, 97.3]

amateur station [1]
     A radio station operated by a duly authorized person interested in radio technique solely with a personal aim and without pecuniary interest. [47 USC 153]

amateur station [2]
     A station in an amateur radio service consisting of the apparatus necessary for carrying on radiocommunications. [47 CFR 97.3]

amateur station [3]
     A station in the amateur service. (RR) [47 CFR 2.1]

American sign language (ASL)

A visual language based on hand shape, position, movement, and orientation of the hands in relation to each other and the body. [47 CFR 64.601]

amortization

The systematic recoveries, through ratable charges to expense, of the cost of assets. [47 CFR 32.9000]

amplitude modulated stage

The radiofrequency stage to which the modulator is coupled and in which the carrier wave is modulated in accordance with the system of amplitude modulation and the characteristics of the modulating wave. [47 CFR 73.14]

amplitude modulation (AM)

A system of modulation in which the envelope of the transmitted wave contains a component similar to the wave form of the signal to be transmitted. [47 CFR 73.681]

amplitude modulator stage

The last amplifier stage of the modulating wave amplitude modulates a radio-frequency stage. [47 CFR 73.14]

AMTS

Automated maritime telecommunications systems [47 CFR 80.5, 97.303]

AMVER

Automated mutual-assistance vessel rescue system [47 CFR 80.5]

ancillary agreement

Any agreement relating to the dismissal of an application or settling of a proceeding, including any agreement on the part of an applicant or principal of an applicant to render consulting services to another party or principal of another party in the proceeding. [47 CFR 73.3526]

ANI (automatic number identification)
Refers to the delivery of the calling party's billing number by a local exchange carrier to any interconnecting carrier for billing or routing purposes, and to the subsequent delivery of such number to end users. [47 CFR 64.1600]

annual revenue requirement
The sum of the return component and the expense component. [47 CFR 69.2]

ANSI
American National Standards Institute [47 CFR 1.1307]

antenna [1]
A device that converts radio frequency electrical energy to radiated electromagnetic energy and vice versa; in a transmitting station, the device from which radio waves are emitted. [47 CFR 22.99]

antenna [2]
The radiating system (for transmitting, receiving or both) and the structure holding it up (tower, pole or mast). It also means everything else attached to the radiating system and the structure. [47 CFR 95.208]

antenna current
The radio-frequency current in the antenna with no modulation. [47 CFR 73.14]

antenna electrical beam tilt
The shaping of the radiation pattern in the vertical plane of a transmitting antenna by electrical means so that maximum radiation occurs at an angle below the horizontal plane.[47 CFR 73.681]

antenna farm
An area that defined as a geographical location, with established boundaries, designated by the Federal Communications Commission, in which antenna towers with a common impact on aviation may be grouped. [47 CFR 17.2]

antenna frequency
Center of a frequency band assigned to a station. [47 CFR 90.7]

antenna height above average terrain (HAAT) [1]

HAAT is calculated by: determining the average of the antenna heights above the terrain from 3 to 16 kilometers (2 to 10 miles) from the antenna for the eight directions evenly spaced for each 45° of azimuth starting with True North (a different antenna height will be determined in each direction from the antenna): and computing the average of these separate heights. In some cases less than eight directions may be used. (See §73.313(d).) Where circular or elliptical polarization is used, the antenna height above average terrain must be based upon the height of the radiation of the antenna that transmits the horizontal component of radiation. [47 CFR 73.310]

antenna height above average terrain [2]

The average of the antenna heights above the terrain from approximately 3.2 (2 miles) to 16.1 kilometers (10 miles) from the antenna for the eight directions spaced evenly for each 45 degrees of azimuth starting with True North. (In general, a different antenna height will be determined in each direction from the antenna. The average of these various heights is considered the antenna height above the average terrain. In some cases less than 8 directions may be used. See §73.684(d). Where circular or elliptical polarization is employed, the antenna height above average terrain shall be based upon the height of the radiation center of the antenna which transmits the horizontal component of radiation. [47 CFR 73.681]

antenna height above average terrain (HAAT) [3]

Height of the center of the radiating element of the antenna above the average terrain. (See S 90.309(a)(4) for calculation method.) [47 CFR 90.7]

antenna height above sea level

The height of the topmost point of the antenna above mean sea level. [47 CFR 90.7]

antenna input power

The product of the square of the antenna current and the antenna resistance at the point where the current is measured. [47 CFR 73.14]

antenna mechanical beam tilt

The intentional installation of a transmitting antenna so that its axis is not vertical, in order to change the normal angle of maximum radiation in the vertical plane. [47 CFR 73.681]

antenna power gain [1]

The square of the ratio of the root-mean-square free space field intensity produced at one mile in the horizontal plane, in millivolts per meter for one kilowatt antenna input power to 137.6 mV/m. This ratio should be expressed in decibels (dB). (If specified for a particular direction, antenna power gain is based on the field strength in that direction only.) [47 CFR 21.2]

antenna power gain [2]

The square of the ratio of the root-mean-square free space field strength produced at 1 kilometer in the horizontal plane, in millivolts per meter for one kW antenna input power to 221.4 mV/m. This ratio should be expressed in decibels (dB). (If specified for a particular direction, antenna power gain is based on the field strength in that direction only.) [47 CFR 73.310, 73.681]

antenna power input

The radio frequency peak or RMS power, as the case may be, supplied to the antenna from the antenna transmission line and its associated impedance matching network. [47 CFR 21.2, 94.3]

antenna resistance

The total resistance of the transmitting antenna system at the operating frequency and at the point at which the antenna current is measured. [47 CFR 73.14]

antenna structure [1]

A structure comprising an antenna, the tower or other structure that exists solely to support antennas, and any surmounting appurtenances (attachments such as beacons or lightning rods). [47 CFR 22.99]

antenna structure [2]

Structure on which an antenna is mounted. [47 CFR 90.7]

antenna structures [1]

Includes the radiating and/or receive system, its supporting structures and any appurtenances mounted thereon. [47 CFR 17.2]

antenna structures [2]

The antenna, its supporting structure and anything attached to it. [47 CFR 21.2]

antitrust laws

Has the meaning given it in subsection (a) of the first section of the Clayton Act (15 USC §12(a)), except that such term includes section 5 of the Federal Trade Commission Act (15 USC §45) to the extent that such section 5 applies to unfair methods of competition. [47 USC 303c]

any person aggrieved

Shall include any person with proprietary rights in the intercepted communication by wire or radio, including wholesale or retail distributors of satellite cable programming, and, in the case of a violation of paragraph (4) of subsection (e), shall also include any person engaged in the lawful manufacture, distribution, or sale of equipment necessary to authorize or receive satellite cable programming. [47 USC 605]

AP

Associated Press [47 CFR 11.43]

API

Actual price index [47 CFR 61.3]

applicant

Shall include the entity submitting a short-form application to participate in an auction (FCC Form 175), as well as all holders of partnership and other ownership interests and any stock interest amounting to 5 percent or more of the entity, or outstanding stock, or outstanding voting stock of the entity submitting a short-form application, and all officers and directors of that entity. [47 CFR 1.2105]

appropriate agency official or designee

The Managing Director of the Commission or such other official as may be named by the Managing Director. [47 CFR 1.1901]

archival quality microfiche
A silver halide master microfiche or a copy made on silver halide film. [47 CFR 22.99]

area of reception
Any geographic area smaller than a zone of reception in which the reception of particular programs is specifically intended and in which broadcast coverage is contemplated, such areas being indicated by countries or parts of countries. [47 CFR 73.701]

area served by cable system
An area actually passed by a cable system and which can be connected for a standard connection fee. [47 CFR 76.1000]

ASCII
An acronym for American Standard Code for Information Interexchange which employs an eight bit code and can operate at any standard transmission baud rate including 300, 1200, 2400, and higher. [47 CFR 64.601]

ASL
American Sign Language [47 C.F.R. 64.601]

aspect ratio
The ratio of picture width to picture height as transmitted. [47 CFR 73.681]

ASSB
Amplitude compandored single sideband modulation [47 CFR 22.161]

assigned frequency [1]
The center of the frequency band assigned to a station. (RR) [47 CFR 2.1, 24.5, 26.4]

assigned frequency [2]
The frequency coinciding with the center of the radio frequency channel in which the station is authorized to work. This frequency does not necessarily correspond to any frequency in an emission. [47 CFR 21.2]

assigned frequency [3]

The frequency coinciding with the center of an authorized bandwidth of emission. [47 CFR 23.1]

assigned frequency band [1]

The frequency band within which the emission of a station is authorized; the width of the band equals the necessary bandwidth plus twice the absolute value of the frequency tolerance. Where space stations are concerned, the assigned frequency band includes twice the maximum Doppler shift that may occur in relation to any point of the earth's surface. (RR) [47 CFR 2.1]

assigned frequency band [2]

The frequency band the center of which coincides with the frequency assigned to the station and the width of which equals the necessary bandwidth plus twice the absolute value of the frequency tolerance. [47 CFR 90.7]

assignment

The designation of priority levels(s) for a defined NSEP telecommunications service for a specified time period. [47 CFR 64.1604, 47 CFR Pt. 216, App.]

assignment (of a radio frequency or radio frequency channel)

Authorization given by an administration for a radio station to use a radio frequency or radio frequency channel under specified conditions. (RR) [47 CFR 2.1]

assignment of authorization

A transfer of a Public Mobile Services authorization from one party to another, voluntarily or involuntarily, directly or indirectly, or by transfer of control of the licensee. [47 CFR 22.99]

assist in intercepting or receiving

Shall include the manufacture or distribution of equipment intended by the manufacturer or distributor (as the case may be) for unauthorized reception or any communications service offered over a cable system in violation of subparagraph (1). [47 USC 553]

Assistant Attorney General

The Assistant Attorney General, Civil Rights Division, United States Department of Justice. [47 CFR 1.1402]

associated broadcasting station(s)

The broadcasting station or stations with which a remote pickup broadcast station or system is licensed as an auxiliary and with which it is principally used. [47 CFR 74.401]

associated equipment

That equipment which functions with a specific type of plant or with two (2) or more types of plant, *e.g.*, switching equipment, network power equipment, circuit equipment, common channel network signaling equipment or network operations equipment. Associated equipment shall be classified to the account appropriate for the type of equipment with which it is predominately used rather than on its own characteristics.

Illustrative examples of associated equipment are alarm and signal apparatus, auxiliary framing, cable and cable racks, distributing frames and equipment thereon, frame and aisle lighting equipment (not permanently attached to the building), relay racks and panels [47 CFR 32.9000]

associated ship unit

A portable VHF transmitter for use in the vicinity of the ship station with which it is associated. [47 CFR 80.5]

associated television broadcast station

A television broadcast station licensed to the licensee of the television auxiliary broadcast station and with which the television auxiliary station is licensed as an auxiliary facility. [47 CFR 74.631]

association

The telephone company association described in subpart G of this part. [47 CFR 69.2]

asynchronous devices

Devices that transmit RF energy at irregular time intervals, as typified by local area network data systems. [47 CFR 15.303]

AT&T

American Telephone and Telegraph [47 CFR 11.43]

AT&T Consent Decree

The order entered August 24, 1982, in the antitrust action styled United States v. Western Electric, Civil Action No. 82-0192, in the United States District Court for the District of Columbia, and includes any judgment or order with respect to such action entered on or after August 24, 1982. [47 USC 153]

ATIS

Automatic terminal information services [47 CFR 2.106]

ATIS

Automatic transmitter identification system [47 CFR 25.308]

attended operation

Operation of a station by a qualified operator on duty at the place where the transmitting apparatus is located with the transmitter in plain view of the operator. [47 CFR 74.901, 78.5]

attributable interest

For purposes of determining whether a party has an "attributable interest" as used in this subpart, the definitions contained in the notes to 76.501 shall be used, provided, however that (1) the single majority shareholder provisions of Note 2 (b) to 76.501 and the limited partner insulation provisions of Note 2(g) to 76.501 shall not apply; and (2) the provisions of Note 2(a) to 76.501 regarding five (5) percent interests shall include all voting or nonvoting stock or limited partnership equity interests of five (5) percent or more. [47 CFR 76.1000]

attributable interests

Partnership and other ownership interests and any stock interest amounting to 20 percent or more of the equity, or outstanding stock, or outstanding voting stock of a licensee or applicant will be attributable. [47 CFR 90.814]

audio programming services

Programming provided by, or generally considered to be comparable to programming provided by, a radio broadcast station. [47 USC 271]

audit

A quality assurance review in response to identified problems. [47 CFR 64.1604, 47 CFR Pt. 216, App.]

auditory assistance device

An intentional radiator used to provide auditory assistance to a handicapped person or persons. Such a device may be used for auricular training in an education institution, for auditory assistance at places of public gatherings, such as a church, theater, or auditorium, and for auditory assistance to handicapped individuals, only, in other locations. [47 CFR 15.3]

aural broadcast intercity relay (ICR) station

A fixed station for the transmission of aural program material between radio broadcast stations, other than international broadcast stations, between FM radio broadcast stations and their co-owned FM booster stations, between noncommercial educational FM radio stations and their co-owned noncommercial educational FM translator stations assigned to reserved channels (Channels 201 to 220), between FM radio stations and FM translator stations operating within the coverage contour of their primary stations, or for such other purposes as authorized in §74.531. [47 CFR 74.501]

aural broadcast microwave booster station

A fixed station in the broadcast auxiliary service that receives and amplifies signals of an aural broadcast STL or intercity relay station and retransmits them on the same frequency. [47 CFR 74.501]

aural broadcast STL station

A fixed station for the transmission of aural program material between the studio and the transmitter of a broadcasting station other than an international broadcasting station. [47 CFR 74.501]

aural center frequency

(1) The average frequency of the emitted wave when modulated by a sinusoidal signal; (2) The frequency of the emitted wave without modulation. [47 CFR 73.681]

aural transmitter

The radio equipment for the transmission of the aural signal only. [47 CFR 73.681]

authorization

A written instrument or oral statement issued by the FCC conveying authority to operate, for a specified term, a station in the Public Mobile Services. [47 CFR 22.99]

authorized bandwidth [1]

The maximum width of the band of frequencies permitted to be used by a station. This is normally considered to be the necessary or occupied bandwidth, whichever is greater. [47 CFR 21.2, 24.5, 26.4]

authorized bandwidth [2]

The occupied or necessary bandwidth, whichever is greater, authorized to be used by a station. [47 CFR 22.99, 74.401]

authorized bandwidth [3]

The maximum bandwidth authorized to be used by a station as specified in the station license. This shall be occupied bandwidth or necessary bandwidth, whichever is greater. [47 CFR 23.1]

authorized bandwidth [4]

The maximum bandwidth authorized to be used by a station as specified in the station license. (See §§2.202 and 78.104.) [47 CFR 78.5]

authorized bandwidth [5]

The frequency band, specified in kilohertz and centered on the carrier frequency containing those frequencies upon which a total of 99 percent of the radiated power appears, extended to include any discrete frequency upon which the power is at least 0.25 percent of the total radiated power. [47 CFR 90.7]

authorized bandwidth [6]

Maximum permissible bandwidth of a transmission. [47 CFR 95.669]

authorized billing agent

A third party hired by a telecommunications service provider to preform billing and collection services for the telecommunications service provider. [47 CFR 64.1201]

authorized carrier [1]

(1) Except as provided in paragraph (b)(2) of this section, the term "authorized carrier" means a communications common carrier which is authorized by the Federal Communications Commission under the Communications Act of 1934, as amended, to provide services by means of communications satellites.

(2) For the purposes of Subpart H of this part, the term "authorized carrier" means a communications common carrier which is specifically authorized or which is a member of a class of carriers authorized by the Commission to own shares of stock in the corporation. [47 CFR 25.103]

authorized carrier [2]

For the purposes of this subpart, the term "authorized carrier" means a communications common carrier which is specifically authorized or which is a member of a class of carriers authorized by the Commission to own shares of stock in the corporation. [47 CFR 25.502]

authorized frequency

The frequency assigned to a station by the Commission and specified in the instrument of authorization. [47 CFR 5.4, 21.2]

authorized power [1]

The power assigned to a radio station by the Commission and specified in the instrument of authorization. The authorized power does not necessarily correspond to the power used by the Commission for purposed of its Master Frequency Record (MFR) and notification to the International Telecommunication Union. [47 CFR 5.4]

authorized power [2]

The maximum power a station is permitted to use. This power is specified by the Commission in the station's authorization. [47 CFR 21.2]

authorized reference frequency

A frequency having a fixed and specific position with respect to the assigned frequency. [47 CFR 23.1]

authorized service

The term "authorized service" of a point-to-point radiotelegraph or radiotelephone station means the transmission of public correspondence to a point of communication as defined herein subject to such special provisions as may be contained in the license of the station or in accordance with §23.53. [47 CFR 23.1]

authorized spectrum

The spectral width of that portion of the electromagnetic spectrum within which the emission power of the authorized transmitter(s) must be contained, in accordance with the rules in this part. The authorized spectrum comprises one channel bandwidth or the bandwidths of two or more contiguous channels. [47 CFR 22.99]

autodialer (see automatic telephone dialing system)

automated maritime telecommunications system (AMTS)

An automatic, integrated and interconnected maritime communications system. [47 CFR 80.5]

automated mutual-assistance vessel rescue system (AMVER)

An international system, operated by the U.S. Coast Guard, which provides aid to the development and coordination of search and rescue (SAR) efforts. Data is made available to recognized SAR agencies or vessels of any nation for reasons related to marine safety. [47 CFR 80.5]

automatic control

The use of devices and procedures for control of a station when it is transmitting so that compliance with the FCC Rules is achieved without the control operator being present at a control point. [47 CFR 97.3]

automatic number identification (see ANI)

automatic relay station

A remote pickup broadcast base station which is actuated by automatic means and is used to relay transmissions between remote pickup broadcast base and mobile stations, between remote pickup broadcast mobile stations and from remote pickup broadcast mobile stations to broadcasting stations. (Automatic operation is not operation by remote control.) [47 CFR 74.401]

automatic telephone dialing system (autodialer)

Equipment which has the capacity to store or produce telephone numbers to be called using a random or sequential number generator and to dial such numbers. [47 USC 227, 47 CFR 64.1200]

automatic weather observation station

A land station located at an airport and used to automatically transmit weather information to aircraft. [47 CFR 87.5]

auxiliary aids

Services or devices that enable persons with impaired sensory, manual, or speaking skills to have an equal opportunity to participate in, and enjoy the benefits of, programs or activities conducted by the Commission. For example, auxiliary aids useful for persons with impaired vision include readers, Brailled materials, audio recordings, and other similar services and devices. Auxiliary aids useful for persons with impaired hearing include telephone handset amplifiers, telephones compatible with hearing aids, telecommunication devices for deaf persons (TDD's), interpreters, note takers, written materials, and other similar services and devices. [47 CFR 1.1803]

auxiliary leads

Terminal equipment leads at the interface, other than telephone connections and leads otherwise defined in these Rules, which leads are to be connected either to common equipment or to circuits extending to central office equipment. [47 CFR 68.3]

auxiliary station

An amateur station, other than in a message forwarding system, that is transmitting communications point-to-point within a system of cooperating amateur stations. [47 CFR 97.3]

auxiliary test transmitter

A fixed transmitter used to test Public Mobile systems. [47 CFR 22.99]

average ship station antenna

An actual antenna installed on board ship having a capacitance of 750 picofarads and an effective resistance of 4 ohms at a frequency of 500 kHz, or an artificial antenna having the same electrical characteristics. [47 CFR 80.251]

average terrain [1]

The average elevation of terrain between 3 and 16 kilometers from the antenna site. [47 CFR 24.5, 26.4]

average terrain [2]

The average elevation of terrain between 3.2 and 16 km (2 and 10 miles) from the antenna site. [47 CFR 90.7]

aviation services

Radio-communication services for the operation of aircraft. These services include aeronautical fixed service, aeronautical mobile service, aeronautical radiodetermination service, and secondarily, the handling of public correspondence on frequencies in the maritime mobile and maritime mobile satellite services to and from aircraft. [47 CFR 87.5]

aviation service organization

Any business firm which maintain facilities at an airport for the purposes of one or more of the following general aviation activities: (a) aircraft fueling; (b) aircraft services (*e.g.* parking, storage, tie-downs); (c) aircraft maintenance or sales; (d) electronics equipment maintenance or sales; (e) aircraft rental, air taxi service or flight instructions; and (f) baggage and cargo handling, and other passenger or freight services. [47 CFR 87.5]

aviation support station

An aeronautical station used to coordinate aviation services with aircraft and to communicate with aircraft engaged in unique or specialized activities. (See Subpart K) [47 CFR 87.5]

AVM

Automatic vehicle monitoring [47 CFR 90.231]

AWOS

Automatic weather observation services [47 CFR 2.106]

B

band

A zone of pricing flexibility for a service category, which zone is calculated pursuant to §61.47. [47 CFR 61.3]

bandwidth

The width of a frequency band outside of which the mean power of the transmitted signal is attenuated at least 26 dB below the mean power of the transmitted signal within the band. [47 CFR 97.3]

bandwidth occupied by an emission

The band of frequencies comprising 99 percent of the total radiated power extended to include any discrete frequency on which the power is at least 0.25 percent of the total radiated power. [47 CFR 21.2]

baseband

Aural transmitter input signals between 0 and 120 kHz. [47 CFR 73.681]

base earth station

An earth station in the fixed-satellite service or, in some cases, in the land mobile-satellite service, located at a specified fixed point or within a specified area on land to provide a feeder link for the land mobile-satellite service. (RR) [47 CFR 2.1, 25.201]

base period

For carriers subject to §§61.41—61.49, the 12-month period ending six months prior to the effective date of annual price cap tariffs, or for carriers regulated under §61.50, the 24-month period ending six months prior to the effective date of biennial optional incentive plan tariffs. Base year or base period earnings shall not include amounts associated with exogenous adjustments to the PCI for the sharing or lower formula adjustment mechanisms. [47 CFR 61.3]

base row
> The bottom row of a roll-up display. The cursor always remains on the base row. Rows of text roll upwards into the contiguous rows immediately above the base row. [47 CFR 15.119]

base station [1]
> A land station in the land mobile service. (RR) [47 CFR 2.1]

base station [2]
> A station at a specified site authorized to communicate with mobile stations. [47 CFR 90.7]

base transmitter
> A stationary transmitter that provides radio telecommunications service to mobile and/or fixed receivers, including those associated with mobile stations. [47 CFR 22.99]

basic cable service
> Any service tier which includes the retransmission of local television broadcast signals. [47 USC 522]

basic service area
> The minimum specified calling area for which a tariff is prescribed. [47 CFR 32.9000]

basic service elements
> Optional unbundled features that enhanced service providers may require or find useful in the provision of enhanced services, as defined in Amendments of part 69 of the Commission's rules relating to the Creation of Access Charge Subelements for Open Network Architecture, Report and Order, 6 FCC Rcd____, CC Docket No. 89-79, FCC 91-186 (1991). [47 CFR 69.2]

basic telephone service
> Any wireline telephone exchange service, or wireline telephone exchange service facility, provided by a Bell operating company in a telephone

exchange area, except that such term does not include (A) a competitive wireline telephone exchange service provided in a telephone exchange area where another entity provides a wireline telephone exchange service that was provided on January 1, 1984, or (B) a commercial mobile service. [47 USC 274]

basic telephone service information

Network and customer information of a Bell operating company and other information acquired by a Bell operating company as a result of its engaging in the provision of basic telephone service. [47 USC 274]

basic trading area (BTA)

The geographic areas by which the Multipoint Distribution Service is licensed. BTA boundaries are based on the Rand McNally *1992 Commercial Atlas and Marketing Guide*, 123rd Edition, pp. 36-39, and include six additional BTA-like areas as specified in §21.924(b). [47 CFR 21.2]

basic trading areas (BTAs)

Service areas that are based on the Rand McNally *1992 Commercial Atlas & Marketing Guide*, 123rd Edition, at pages 38-39, with the following additions licensed separately as BTA-like areas: American Samoa; Guam, Northern Mariana Islands; Mayaguez/Aguadilla-Ponce, Puerto Rico; San Juan, Puerto Rico; and the Untied States Virgin Islands. The Mayaguez/Aguadilla-Ponce BTA-like service area consists of the following municipios: Adjuntas, Aguada, Aguadilla, Anasco, Arroyo, Cabo Rojo, Coamo, Guanica, Guayama, Guayanilla, Hormigueros, Isabela, Jayuya, Juana Diaz, Lajas, Las Marias, Maricao, Maunabo, Mayaguez, Moca, Patillas, Penuelas, Ponce, Quebradillas, Rincon, Sabana Grande, Salinas, San German, Santa Isabel, Villalba, and Yauco. The San Juan BTA-like service area consists of all other municipios in Puerto Rico. [47 CFR 90.7]

basket

Any class or category of tariffed services (1) which is established by the Commission pursuant to price cap regulation; (2) the rates of which are reflected in an Actual Price Index; and (3) the related costs of which are reflected in a Price Cap Index. [47 CFR 61.3]

baudot

A seven bit code, only five of which are information bits. Baudot is used by some text telephones to communicate with each other at a 45.5 baud rate. [47 CFR 64.601]

beacon

An amateur station transmitting communications for the purposes of observation of propagation and reception or other related experimental activities. [47 CFR 97.3]

Bell operating company [1]

(A) Any of the following companies: Bell Telephone Company of Nevada, Illinois Bell Telephone Company, Indiana Bell Telephone Company, Incorporated, Michigan Bell Telephone Company, New England Telephone and Telegraph Company, New Jersey Bell Telephone Company, New York Telephone Company, U. S. West Communications Company, South Central Bell Telephone Company, Southern Bell Telephone Company and Telegraph Company, The Chesapeake and Potomac Telephone Company, The Chesapeake and Potomac Telephone Company of Maryland, The Chesapeake and Potomac Telephone Company of Virginia, the Chesapeake and Potomac Telephone Company of West Virginia, The Diamond State Telephone Company, The Ohio Bell Telephone Company, The Pacific Telephone and Telegraph Company, or Wisconsin Telephone Company, and

(B) includes any successor or assign of any such company that provides wireline telephone exchange service, but

(C) does not include an affiliate of any such company, other than an affiliate described in subparagraph (A) or (B). [47 USC 153]

Bell operating company [2]

Has the meaning provided in Section 3 [47 USC 153], except that such term includes any entity or corporation that is owned or controlled by such a company (as so defined) but does not include an electronic publishing joint venture owned by such an entity or corporation. [47 USC 274]

BETRS

Basic exchange telephone radio systems [47 CFR 22.702]

bids or bidding strategies

Shall include capital calls or requests for additional funds in support of bids or bidding strategies. [47 CFR 1.2105]

big three expense factors

The ratios of the sum of Big Three Expenses apportioned to each element or category to the combined Big Three Expenses. [47 CFR 69.2]

big three expenses

The combined expense groups comprising: Plant Specific Operations Expense, Accounts 6110, 6120, 6220, 6230, 6310 and 6410; Plant Nonspecific Operations Expenses, Accounts 6510, 6530 and 6540, and Customer Operations Expenses, Accounts 6610 and 6620. [47 CFR 69.2]

billing name and address

The name and address provided to a local exchange company by each of its local exchange customers to which the local exchange company directs bills for its services. [47 CFR 64.1201]

biomedical telemetry device

An intentional radiator used to transmit measurements of either human or animal biomedical phenomena to a receiver. [47 CFR 15.3]

bit rate

The rate of transmission of information in binary (two state) form in bits per unit time. [47 CFR 21.2, 94.3]

blanketing

The interference which is caused by the presence of an AM broadcast signal of one volt per meter (V/m) or greater strengths in the area adjacent to the antenna of the transmitting station. The 1 V/m contour is referred to as the blanket contour and the area within this contour is referred to as the blanket area. [47 CFR 73.14]

blanketing interference

Disturbance in consumer receivers located in the immediate vicinity of a transmitter, caused by currents directly induced into the consumer receiver's circuitry by the relatively high field strength of the transmitter. [47 CFR 22.99]

blanking level

The level of the signal during the blanking interval, except the interval during the scanning synchronizing pulse and the chrominance subcarrier synchronizing burst. [47 CFR 73.681]

B$_n$

bandwidth [47 CFR 2.202]

BOCs

Bell operating companies [47 CFR 22.903]

book cost [1]

The amount at which property is recorded in these accounts, without deduction of related allowances. [47 CFR 32.9000]

book cost [2]

The cost of property as recorded on the books of a company. [47 CFR Pt. 36, App.]

box

The area surrounding the active character display. In Text Mode, the box is the entire screen area defined for display, whether or not displayable characters appear. In Caption Mode, the box is dynamically redefined by each caption and each element of displayable characters within a caption. The box (or boxes, in the case of a multiple-element caption) includes all the cells of the displayed characters, the nontransparent spaces between them, and one cell at the beginning and end of each row within a caption element in those decoders that use a solid space to improve legibility. [47 CFR 15.119]

bridge to bridge station

A radio station located on a ship's navigational bridge or main control station operating on a specified frequency which is used only for navigational communications, in the 156-162 MHZ band. [47 CFR 80.5]

bridged

A bridged connection is a parallel connection. [47 CFR 68.502]

broadband PCS

PCS services operating in the 1850-1890 MHZ, 1930-1970 MHZ, 2130-2150 MHZ, and 2180-2200 MHZ bands. [47 CFR 24.5]

broadcast day

That period of time between the station's sign-on and its sign-off. [47 CFR 73.1700]

broadcast network-entity

A broadcast network-entity is an organization which produces programs available for simultaneous transmission by 10 or more affiliated broadcast stations and having distribution facilities or circuits available to such affiliated stations at least 12 hours each day. [47 CFR 74.2]

broadcasting [1]

The dissemination of radio communications intended to be received by the public, directly or by the intermediary of relay stations. [47 USC 153]

broadcasting [2]

Transmissions intended for reception by the general public, either direct or relayed. [47 CFR 97.3]

broadcasting-satellite service

A radiocommunication service in which signals transmitted or retransmitted by space stations are intended for direct reception by the general public. [47 CFR 2.1]

broadcasting service

A radiocommunication service in which the transmissions are intended for direct reception by the general public. This service may include sound transmissions, television transmissions or other types of transmission. (CONV) [47 CFR 2.1]

broadcast station [1]

A radio station equipped to engage in broadcasting as herein defined. [47 USC 153]

broadcasting station [2]

A community antenna television system. [47 USC 315]

broadcasting station [3]

A station in the broadcasting service. (RR) [47 CFR 2.1]

BST

Basic service tier [47 CFR 76.922]

BTA

Basic trading area [47 CFR 21.2]

BTA authorization holder

The individual or entity authorized by the Commission to provide Multipoint Distribution Service to the population of a BTA. [47 CFR 21.2]

BTA service area

The area within the boundaries of a BTA to which a BTA authorization holder may provide Multipoint Distribution Service. This area excludes the protected service areas of incumbent MDS stations and the registered receive sites of previously authorized and proposed ITFS stations. [47 CFR 21.2]

BTSC

Broadcast television systems committee recommendation for multi-channel television sound transmission and audio processing as defined in FCC Bulletin OET 60. [47 CFR 73.681]

build-out transmitters

In the cellular radiotelephone service, transmitters added to the first cellular system authorized on a channel block in a cellular market during the five year build-out period in order to expand the coverage of the system within the market. [47 CFR 22.99]

bulk basis

Billing name and address information for all the local exchange service subscribers of local exchange carrier. [47 CFR 64.1201]

business day

All days, including days when the Commission opens later than the time specified in Rule §0.403, which are not holidays as defined above. [47 CFR 1.4]

business or institutional logogram

Any aural or visual letters or words, or any symbol or sign, which is used for the exclusive purpose of identifying any corporation, company, or other organization, and which is not used for the purpose of promoting the products, services, or facilities of such corporation, company, or other organization. [47 USC 399a]

business owned by members of minority groups and/or women [1]

An entity (1) that has a control group composed 100 percent of members of minority groups and/or women who are United States Citizens, and (2) such control group owns and holds 50.1 percent of the voting interests, if a corporation, and (i) owns and holds 50.1 percent of the total equity in the entity, provided that all other investors hold passive interests; or (ii) holds 25 percent of the total equity in the entity, provided that no single other investor holds more than 25 percent passive equity interests in the entity. In a partnership, all general partners must be members of minority groups and/or women. Ownership interests shall be calculated on a fully diluted basis; all agreements such as warrants, stock options and convertible debenture will generally be treated as if the rights thereunder already have been fully exercised, except that such agreements may not be used to appear to terminate or divest ownership interests before they actually do so. [47 CFR 24.320]

business owned by members of minority groups and/or women [2]

An entity (1) in which the qualifying investor members of an applicant's control group are members of minority groups and/or women who are United States citizens; and (2) that complies with the requirements of §24.715 (b)(3) and (b)(5) or §24.715 (b)(4) and (b)(6). [47 CFR 24.720]

business owned by members of minority groups and/or women [3]

A business owned by members of minority groups and/or women in which minorities and/or women who are U.S. citizens control the applicant, have at least 50.1 percent equity ownership and, in the case of a corporate applicant, a 50.1 percent voting interest. For applicants that are partnerships, every general partner either must be a minority and/or woman (or minorities and/or women) who are U.S. citizens and who individually or together own at least 50.1 percent of the partnership equity, or an entity that is 100 percent owned and controlled by minorities and/or women who are U.S. citizens. The interests of minorities and women are to be calculated on a fully-diluted basis; agreements such as stock options and convertible debentures shall be considered to have a present effect on the power to control an entity and shall be treated as if the rights thereunder already have been fully exercised. However, upon a demonstration that options or conversion rights held by non-controlling principals will not deprive the minority and female principals of a substantial financial stake in the venture or impair their rights to control the designated entity, a designated entity may seek a waiver of the requirement that the equity of the minority and female principals must be calculated on a fully-diluted basis. [47 CFR 90.814]

buying groups (or "agent")
For purposes of the definition of a multichannel video programming distributor set forth in paragraph (e) of this section, means an entity representing the interests of more than one entity distributing multichannel video programming that (1) agrees to be financially liable for any fees due pursuant to a satellite cable programming, or satellite broadcast programming, contract which it signs as a contracting party as a representative of its members or whose members, as contracting parties, agree to joint and several liability; and (2) agrees to uniform billing and standardized contract provisions for individual members; and (3) agrees either collectively or individually on reasonable technical quality standards for the individual members of the group. [47 CFR 76.1000, 76.1300]

BZW
Blizzard warning [47 CFR 11.31]

C

CA

Communications assistant [47 CFR 64.601]

cablecasting

Programming (exclusive of broadcast signals) carried on a cable television system. See paragraphs (y), (z) and (aa) (Classes II, III, and IV cable television channels) of this section. [47 CFR 76.5]

cable channel (channel)

A portion of the electromagnetic frequency spectrum which is used in a cable system and which is capable of delivering a television channel (as television channel is defined by the Commission by regulation). [47 USC 522]

cable fill factor

The ratio of cable conductor or cable pair kilometers in use to total cable conductor or cable pair kilometers available in the plant, *e.g.*, the ratio of revenue producing cable pair kilometers in use to total cable pair kilometers in plant. [47 CFR Pt. 36, App.]

cable home wiring

The internal wiring contained within the premises of a subscriber which begins at the demarcation point. Cable home wiring does not include any active elements such as amplifiers, converter or decoder boxes, or remote control units. [47 CFR 76.5]

cable input selector switch

A transfer switch that is intended as a means to alternate between the reception of broadcast signals via connection to an antenna and the reception of cable television service. [47 CFR 15.3]

cable locating equipment

An intentional radiator used intermittently by trained operators to locate buried cables, lines, pipes, and similar structures or elements. Operation entails coupling a radio frequency signal onto the cable, pipes, etc. and using a receiver to detect the location of that structure or element. [47 CFR 15.3]

cable network-entity

A cable network-entity is an organization which produces programs available for simultaneous transmission by cable systems serving a combined total of at least 5,000,000 subscribers and having distribution facilities or circuits available to such affiliated stations or cable systems. [47 CFR 74.2, 78.5]

cable operator [1]

Any person or group of persons (A) who provides cable service over cable system and directly or through one or more affiliates owns a significant interest in such cable system, or (B) who otherwise controls or is responsible for, through any arrangement, the management and operation of such a cable systems. [47 USC 522]

cable operator [2]

In addition to persons within the definition of cable operator in Section 602, any person who (i) is owned or controlled by, or under common ownership or control with a cable operator, and (ii) provides any wire or radio communications service. [47 USC 551]

cable operator [3]

(1) Includes any operator of any satellite master antenna television system, including a system described in Section 602(7)(A) and any multichannel video programming distributor.

(2) Such term does not include any operator of a system which, in the aggregate, serves fewer than 50 subscribers.

(3) In any case in which a cable operator is the owner of a multiple unit dwelling, the requirements of this section shall only apply to such cable operator with respect to its employees who are primarily engaged in cable telecommunications. [47 USC 554]

cable programming service

Any video programming provided over a cable system, regardless of service tier, including installation or rental of equipment used for the receipt of such video programming, other than (A) video programming carried on the basic service tier, and (B) video programming offered on a per channel or per program basis. [47 USC 543]

cable ready consumer electronics equipment

Consumer electronics TV receiving devices, including TV receivers, videocassette recorders and similar devices, that incorporate a tuner capable of receiving television signals and an input terminal intended for receiving cable television service, and are marketed as "cable ready" or "cable compatible." Such equipment shall comply with the technical standards specified in §15.118. [47 CFR 15.3]

cable service

The one-way transmission to subscribers of video programming, or other programming service; and, subscriber interaction, if any, which is required for the selection of such video programming or other programming service [47 USC 153]. For the purposes of this definition, "video programming" is programming provided by, or generally considered comparable to programming provided by, a television broadcast station; and, "other programming service" is information that a cable operator makes available to all subscribers generally. [47 CFR 76.5]

cable system [1]

A facility, consisting of a set of closed transmission paths and associated signal generation, reception, and control equipment that is designed to provide cable service which includes video programming and which is provided to multiple subscribers within a community, but such term does not include (A) a facility that services only to retransmit the television signals of 1 or more television broadcast stations; (B) a facility that serves subscribers without using any public right-of-way; (C) a facility of a common carrier which is subject, in whole or in part, to the provisions of Title II of this Act, except that such facility shall be considered a cable system (other than for purposes of Section 621(c)) to the extent such facility is used in the transmission of video programming directly to subscribers, unless the extent of such use is solely to provide interactive on-demand services; (D) an open video system that complies with section 653 of this title; or (E) any facilities of any electric utility used solely for operating its electric utility systems. [47 USC 153, 47 USC 522]

cable system (or cable television system) [2]

A facility consisting of a set of closed transmission paths and associated signal generation, reception, and control equipment that is designed to provide cable service which includes video programming and which is provided to multiple subscribers within a community, but such term does not include (1) a facility that services only to retransmit the television signals of one or more television broadcast stations; (2) a facility that serves only subscribers in one or more multiple unit dwellings under common ownership, control or management, unless such facility or facilities uses any public right-or-way; (3) a facility of a common carrier which is subject, in whole or in part, to the provisions of the Title II of the Communications Act of 1934, as amended, except that such facility shall be considered a cable system to the extent such facility is used in the transmission of video programming directly to subscribers; or (4) any facilities of any electric utility used solely for operating its electric utility systems. [47 CFR 76.5]

cable system operator

Any person or group of persons (1) who provides cable service over a cable system and directly or through one or more affiliates owns a significant interest in such cable system; or (2) who otherwise controls or is responsible for, through any arrangement, the management and operation of such a cable system. [47 CFR 76.5]

cable system terminal device (CSTD)

A TV interface device that serves, as its primary function, to connect a cable system under part 76 of this chapter to a TV broadcast receiver or other subscriber premise equipment. Any device which functions as a CSTD in one of its operating modes must comply with the technical requirements for such devices when operating in that mode. [47 CFR 15.3]

cable television relay service (CARS) station

A fixed or mobile station used for the transmission of television and related audio signals, signals of standard and FM broadcast stations, signals of instructional television fixed stations, and cablecasting from the point of reception to a terminal point from the point of reception to a terminal point from which the signals are distributed to the public. [47 CFR 78.5]

cable television relay service PICKUP station

A land mobile CARS station used for the transmission of television signals and related communications from the scenes of events occurring at points removed from cable television studios to cable television studios or headends. [47 CFR 78.5]

cable television relay service studio to headend link (SHL) station

A fixed CARS station used for the transmission of television program material and related communications from a cable television studio to the headend of a cable television system. [47 CFR 78.5]

cable television system operator

A cable television operator is defined in §76.5(cc) of the rules. [47 CFR 74.801]

cable and wire facilities

All equipment or facilities that are described as cable and wire facilities in the *Separations Manual* and included in Account 2410. [47 CFR 69.2]

call sign system

The method used to select a call sign for amateur station over-the-air identification purposes. The call sign systems are (i) *sequential call sign system.* The call sign is selected by the FCC from an alphabetized list corresponding to the geographic region of the licensee's mailing address and operator class. The call sign is shown on the license. The FCC will issue public announcements detailing the procedures of the sequential call sign system; and (ii) *vanity call sign system.* The call sign is selected by the FCC from a list of call signs requested by the licensee. The call sign is shown on the license. The FCC will issue public announcements detailing the procedures of the vanity call sign system. [47 CFR 97.3]

call splashing

The transfer of a telephone call from one provider of operator services to another such provider in such a manner that the subsequent provider is unable or unwilling to determine the location of the origination of the call and, because of such inability or unwillingness, is prevented from billing the call on the basis of such location. [47 USC 226, 47 CFR 64.708]

calling card
    An identifying number or code unique to the individual, that is issued to the individual by a common carrier and enables the individual to be charged by means of a phone bill for charges incurred independent of where the call originates. [47 USC 228]

calling party number
    The subscriber line number or the directory number contained in the calling party number parameter of the call set-up message associated with an interstate call on a Signaling System 7 network. [47 CFR 64.1600]

candidate
    An individual who seeks nomination for election, or election, to federal office, whether or not such individual is elected, and an individual shall be deemed to seek nomination for election, or election, if he has (1) taken the action necessary under the law of a state to qualify himself for nomination for election, or election, to federal office, or (2) received contributions or made expenditures, or has given his consent for any other person to receive contributions or make expenditures, with a view to bringing about his nomination for election, or election, to such office. [47 CFR 64.803]

caption window
    The invisible rectangle which defines the top and bottom limits of a roll-up caption. The window can be 2 to 4 rows high. The lowest row of the window is called the base row. [47 CFR 15.119]

cardinal radials
    Eight imaginary straight lines extending radially on the ground from an antenna location in the following azimuths with respect to true North: 0°, 45°, 90°, 135°, 180°, 225°, 270°, 315°. [47 CFR 22.99]

cargo ship
    Any ship not a passenger ship. [47 USC 153]

cargo ship safety radiotelegraphy certificate
    A certificate issued after an inspection of a cargo ship radiotelegraph station which complies with the applicable Safety Convention radio requirements. [47 CFR 80.5]

carrier (also common carrier)

In a frequency stabilized system, the sinusoidal component of a modulated wave whose frequency is independent of the modulating wave; or the output of a transmitter when the modulating wave is made zero; or a wave generated at a point in the transmitting system and subsequently modulated by the signal; or a wave generated locally at the receiving terminal which, when combined with the side bands in a suitable detector, produces the modulating wave. [47 CFR 21.2, 94.3]

carrier-amplitude regulation (carrier shift)

The change in amplitude of the carrier wave in an amplitude-modulated transmitter when modulation is applied under conditions of symmetrical modulation. [47 CFR 73.14]

carrier cable and wire facilities

All cable and wire facilities that are not subscriber line cable and wire facilities. [47 CFR 69.2]

carrier current system

A system, or part of a system, that transmits radio frequency energy by conduction over the electric power lines. A carrier current system can be designed such that the signals are received by conduction directly from connection to the electric power lines (unintentional radiator) or the signals are received over-the-air due to radiation of the radio frequency signals from the electric power lines (intentional radiator). [47 CFR 15.3]

carrier frequency [1]

The output of a transmitter when the modulating wave is made zero. [47 CFR 21.2]

carrier frequency [2]

The frequency of the unmodulated electrical wave at the output of an amplitude modulated (AM), frequency modulated (FM) or phase modulated (PM) transmitter. [47 CFR 22.99]

carrier frequency [3]

The frequency of an unmodulated electromagnetic wave. [47 CFR 90.7]

carrier power [1]

The average power supplied to the antenna transmission line by a transmitter during one radio frequency cycle under conditions of no modulation. [47 CFR 2.1, 73.701]

carrier power [2]

The average power at the output terminals of a transmitter (other than a transmitter having a suppressed, reduced or controlled carrier) during one radio frequency cycle under conditions of no modulation. [47 CFR 74.401]

carrier power [3]

Average TP during one unmodulated RF cycle. [47 CFR 95.669]

CARS

Community antenna relay stations [47 CFR 2.106, 74.602]

catastrophe

A disaster or imminent disaster involving violent or sudden event affecting the public. [47 CFR 73.1225]

categories of ships

(1) When referenced in Part II of Title III of the Communications Act or the radio provisions of the Safety Convention, a ship is a *passenger ship* if it carries or is licensed or certificated to carry more than twelve passengers. A *cargo ship* is any ship not a passenger ship.

(2) A *commercial transport vessel* is any ship which is used primarily in commerce (i) for transporting persons or goods to or from any harbor(s) or port(s) or between places within a harbor or port area, or (ii) in connection with the construction, change in construction, servicing, maintenance, repair, loading, unloading, movement, piloting, or salvaging of any other ship or vessel.

(3) The term *passenger carrying vessel*, when used in reference to Part III, Title III of the Communications Act of the Great Lakes Radio Agreement, means any ship transporting more than six passengers for hire.

(4) *Power-driven vessel.* Any ship propelled by machinery.

(5) *Towing vessel.* Any commercial ship engaged in towing another ship astern, alongside or by pushing ahead.

(6) *Compulsory ship.* Any ship which is required to be equipped with radiotelecommunication equipment in order to comply with the radio or radio-navigation provisions of a treaty or statute to which the vessel is subject.

(7) *Voluntary ship.* Any ship which is not required by treaty or statute to be equipped with radiotelecommunication equipment. [47 CFR 80.5]

category

A grouping of items of property or expense to facilitate the apportionment of their costs among the operations and to which, ordinarily, a common measure of use is applicable. [47 CFR Pt. 36, App.]

CATV

Cable television [47 CFR 1.1404]

CB

Citizens band [47 CFR 73.1207]

CB receiver

Any receiver that operates in the Personal Radio Services on frequencies allocated for Citizens Band (CB) Radio Service stations, as well as any receiver provided with a separate band specifically designed to receive the transmissions of CB stations in the Personal Radio Services. This includes the following: (1) a CB receiver sold as a separate unit of equipment; (2) the receive section of a CB transceiver; (3) a converter to be used with any receiver for the purpose of receiving CB transmissions; and, (4) a multiband receiver that includes a band labeled "CB" or "11-meter" in which such band can be separately selected, except that an Amateur Radio Service receiver that was manufactured prior to January 1, 1960, and which includes and 11-meter band shall not be considered a CB receiver. [47 CFR 15.3]

CBS

Columbia Broadcasting System [47 CFR 11.43]

CB transmitter

A transmitter than operates or is intended to operate at a station authorized in the CB. [47 CFR 95.669]

CCD

Census County Division [47 CFR 73.525]

CCIR
> International Radio Consultative Committee [47 CFR 2.202, 80.5]

CCITT
> International Telegraph and Telephone Consultative Committee [47 CFR 80.1101]

CCT
> Closed circuit test [47 CFR 11.62]

CEA
> Component economic area [47 CFR 26.102]

cell [1]
> The discrete screen area in which each displayable character or space may appear. A cell is one row high and column wide. [47 CFR 15.119]

cell [2]
> The service area of an individual transmitter location in a cellular system. [47 CFR 22.99]

cellular geographic service area
> The geographic area served by a cellular system, within which that system is entitled to protection and adverse effects are recognized, for the purpose of determining whether a petitioner has standing. See §22.911. [47 CFR 22.99]

cellular markets
> Standard geographic areas used by the FCC for administrative convenience in the licensing of cellular systems. See §22.909. [47 CFR 22.99]

cellular radiotelephone service
> A radio service in which common carriers are authorized to offer and provide cellular service for hire to the general public. This service was formerly titled Domestic Public Cellular Radio Telecommunications Service. [47 CFR 22.99]

cellular repeater

In the Cellular Radiotelephone Service, a stationary transmitter or device that automatically re-radiates the transmissions of base transmitters at a particular cell site and mobile stations communicating with those base transmitters, with or without channel translation. [47 CFR 22.99]

cellular service

Radio telecommunication services provided using a cellular system. [47 CFR 22.99]

cellular system

An automated high-capacity system of one or more multichannel base stations designed to provide radio telecommunication services to mobile stations over a wide area in a spectrally efficient manner. Cellular systems employ techniques such as low transmitting power and automatic hand-off between base stations of communications in progress to enable channels to be reused at relatively short distances. Cellular systems may also employ digital techniques such as voice encoding and decoding, data compression, error correction, and time or code division multiple access in order to increase system capacity. [47 CFR 22.99]

CEM

Civil emergency message [47 CFR 11.31]

center frequency [1]

The frequency of the middle of the bandwidth of a channel. [47 CFR 22.99]

center frequency [2]

(1) The average frequency of the emitted wave when modulated by a sinusoidal signal. (2) The frequency of the emitted wave without modulation. [47 CFR 73.310]

central office [1]

A landline termination center used for switching and interconnection of public message communication circuits. [47 CFR 21.2]

central office [2]

A switching unit, in a telephone system which provides service to the general public, having the necessary equipment and operations arrangements for terminating and interconnecting subscriber lines and trunks or trunks only. There may be more than one central office in a building. [47 CFR Pt. 36, App.]

central office equipment (COE)

All equipment or facilities that are described as Central Office Equipment in the *Separations Manual* and included in Accounts 2210, 2220 and 2230. [47 CFR 69.2]

central office transmitter

A fixed transmitter in the Rural Radiotelephone Service that provides service to rural subscriber stations. [47 CFR 22.99]

certification

Any technical process whereby a party determines whether a product, for use by more than one local exchange carrier, conforms with the specified requirements pertaining to such product. [47 USC 273]

CFR

Code of Federal Regulations [47 CFR 0.411]

CGSA

Cellular geographic service area. [47 CFR 22.123]

chain broadcasting

Simultaneous broadcasting of an identical program by two or more connected stations. [47 USC 153]

change in rate structure

A restructuring or other alternation of the rate components for an existing service. [47 CFR 61.3]

channel [1]

The portion of the electro-magnetic spectrum assigned by the FCC for one emission. In certain circumstances, however, more than one emission may be transmitted on a channel. See, for example, §22.161. [47 CFR 22.99]

channel [2]

An electrical path suitable for the transmission of communications between two or more points, ordinarily between two or more stations or between channel terminations in Telecommunication Company central offices. A channel may be furnished by wire, fiber optics, radio or a combination thereof. [47 CFR Pt. 36, App.]

channel bandwidth

The spectral width of a channel, as specified in this part, within which 99% of the emission power must be contained. [47 CFR 2.99]

channel block

A group of channels that are assigned together, not individually. [47 CFR 22.99]

channel equipment

Equipment in the private line channel of the telephone network that furnishes telephone tip and ring, telephone tip 1 and ring 1, and other auxiliary or supervisory signaling leads for connection at the private line channel interface (where tip 1 and ring 1 is the receive pair for 4-wire telephone connections.). [47 CFR 68.3]

channel frequencies

Reference frequencies from which the carrier frequency, suppressed or otherwise, may not deviate by more than the specified frequency tolerance. [47 CFR 95.669]

channel loading

The number of mobile transmitters authorized to operate on a particular channel within the same service area. [47 CFR 90.7]

channel pair

Two channels that are assigned together, not individually. In this part, channel pairs are indicated by an ellipsis between the center frequencies. [47 CFR 22.99]

characteristic frequency

A frequency which can be easily identified and measured in a given emission. Note: A carrier frequency may, for example, be designated as the characteristic frequency. [47 CFR 2.1]

charge number

The delivery of the calling party's billing number in a Signaling System 7 environment by a local exchange carrier to any interconnecting carrier for billing or routing purposes, and to the subsequent delivery of such number to end users. [47 CFR 64.1600]

charges

The price for service based on tariffed rates. [47 CFR 61.3]

children's programming

Programs originally produced and broadcast primarily for an audience of children 12 years old and younger. [47 CFR 73.670]

chrominance

The colorimetric difference between any color and a reference color of equal luminance, the reference color having a specific chromaticity. [47 CFR 73.681]

chrominance subcarrier

The carrier which is modulated by the chrominance information. [47 CFR 73.681]

CIC

Carrier identification code [47 CFR 69.2]

circuit

A fully operative communications path established in the normal circuit layout and currently used for message, WATS access, TWX, or private line services. [47 CFR Pt. 36, App.]

circuit kilometers

The route kilometers or revenue producing circuits in service, determined by measuring the length in terms of kilometers, of the actual path followed by the transmission medium. [47 CFR Pt. 36, App.]

CISPR
International Special Committee on Radio Interference [47 CFR 15.35]

citizens band radio service
Has the meaning given it by the Commission by rule. [47 USC 403]

CIV
Civil authorities [47 CFR 11.31]

Civil Air Patrol station
A station used exclusively for communications of the Civil Air Patrol. [47 CFR 87.5]

claim and debt
Are deemed synonymous and interchangeable. They refer to an amount of money or property which has been determined by an appropriate agency official to be owed to the United States from any person, organization, or entity, except another federal agency. They include amounts owing to the United States on account of loans insured or guaranteed by the United States and all other amounts due the United States from fees, leases, rents, royalties, services, sales of real or personal property, overpayments, fines, penalties, damages interest, taxes, and forfeitures (except those arising under the Uniform Code of Military Justice), and other similar sources. [47 CFR 1.1901]

class A digital device
A digital device that is marketed for use in a commercial, industrial or business environment, exclusive of a device which is marketed for use by the general public or is intended to be used in the home. [47 CFR 15.3]

class B digital device
A digital device that is marketed for use in a residential environment notwithstanding use in commercial, business and industrial environments. Examples of such devices include, but are not limited to, personal computers, calculators, and similar electronic devices that are marketed for use by the general public. [47 CFR 15.3]

class of emission

The set of characteristics of an emission, designated by standard symbols, *e.g.*, type of modulation, modulating signal, type of information to be transmitted, and also if appropriate, any additional signal characteristics. (RR) [47 CFR 2.1]

class I cable television channel

A signaling path provided by a cable television system to relay to subscriber terminals television broadcast programs that are received off-the-air or are obtained by microwave or by direct connection to a television broadcast station. [47 CFR 76.5]

class II cable television channel

A signaling path provided by a cable television system to deliver to subscriber terminals television signals that are intended for reception by a television broadcast receiver without the use of an auxiliary decoding device and which signals are not involved in a broadcast transmission path. [47 CFR 76.5]

class III cable television channel

A signaling path provided by a cable television system to delivery to subscriber terminals signals that are intended for reception by equipment other than a television broadcast receiver or by a television broadcast receiver only when used with auxiliary decoding equipment. [47 CFR 76.5]

class IV cable television channel

A signaling path provided by a cable television system to transmit signals of any type from a subscriber terminal to another point in the cable television system. [47 CFR 76.5]

CMRS

Commercial mobile radio service [47 CFR 20.3]

CMSA

Consolidated metropolitan statistical area [47 CFR 26.102]

CNN

Cable News Network [47 CFR 11.43]

coast earth station
  An earth station in the fixed-satellite service or, in some cases, in the maritime mobile-satellite service, located at a specified fixed point on land to provide a feeder link for the maritime mobile-satellite service. (RR) [47 CFR 2.1]

coast station
  A land station in the maritime mobile service. [47 CFR 2.1, 80.5]

coastal waters
  Waters along the Pacific Coast of Washington State and Vancouver Island within the Canada/U.S.A. Coordination Zone. [47 CFR 80.57]

coastal waters local channel
  Same as inland waters local channel except for technical characteristics. [47 CFR 80.57]

coastal waters primary channel
  Same as inland waters primary channel except for technical characteristics. [47 CFR 80.57]

coastal waters public correspondence sector
  A distinct geographical area in which one primary and one supplementary channel is allotted. Local channels may also be authorized. [47 CFR 80.57]

coastal waters supplementary channel
  Same as inland waters supplementary channel except for technical characteristics. [47 CFR 80.57]

COE
  Central office equipment [47 CFR 69.2]

coin service
  Central office implemented coin telephone service. [47 CFR 68.3]

coin-implemented telephone

A telephone containing all circuitry required to execute coin acceptance and related functions within the instrument itself and not requiring coin service signaling from the central office. [47 CFR 68.3]

COLEM

Commercial operator license examination manager [47 CFR 13.3]

color transmission

The transmission of color television signals which can be reproduced with different values of hue, saturation, and luminance. [47 CFR 73.681]

column

One of 32 vertical divisions of the screen, each of equal width, extending approximately across the full width of the safe caption area as defined in paragraph (n)(12) of this section. Two additional columns, one at the left of the screen and one at the right may be defined for the appearance of a box in those decoders which use a solid space to improve legibility, but no displayable characters may appear in those additional columns. For reference, columns may be numbers 0 to 33, with columns 1 to 32 reserved for displayable characters. [47 CFR 15.119]

combined audio harmonics

The arithmetical sum of the amplitudes of all the separate harmonic components. Root sum square harmonic readings may be accepted under conditions prescribed by the FCC. [47 CFR 73.14]

commercial communications

Communications between coast stations and ship stations aboard commercial transport vessels, or between ship stations aboard commercial transport vessels, which relate directly to the purposes for which the ship is used including the piloting of vessels, movements of vessels, obtaining vessel supplies, and scheduling of repairs. [47 CFR 80.5]

commercial contractor

The commercial firm to whom the Commission annually awards a contract to make copies of Commission records for sale to the public. [47 CFR 61.3]

commercial FM translator

An FM broadcast translator station which rebroadcasts the signals of a commercial FM radio broadcast station. [47 CFR 74.1201]

commercial matter

Air time sold for purposes of selling a product or service. [47 CFR 73.670]

commercial mobile radio service

A mobile service that is (a)(1) provided for profit, *i.e.*, with the intent of receiving compensation or monetary gain; (2) an interconnected service; and (3) available to the public, or to such classes of eligible users as to be effectively available to a substantial portion of the public; or (b) The functional equivalent of such a mobile service described in paragraph (a) of this section. [47 CFR 20.3]

commercial mobile service

Any mobile service (as defined in Section 3) that is provided for profit and makes interconnected service available (A) to the public or (B) to such classes of eligible users as to be effectively available to a substantial portion of the public, a specified by regulation by the Commission. [47 USC 332]

commercial radio operator

A person holding a license or licenses as specified in §13.7(b). [47 CFR 13.3]

commercial use [1]

The provision of video programming, whether or not for profit. [47 USC 532]

commercial use [2]

A request from or on behalf of one who seeks information for a use or purpose that furthers the commercial interests of the requester. In determining whether a requester properly falls within this category, the Commission shall determine the use to which a requester will put the documents requested. Where the commission has reasonable cause to question the use to which a requester will put the documents sought, or where that use is not clear from the request itself, the Commission shall seek additional clarification before assigning the request to a specific category. [47 CFR 0.466]

commercially impracticable

With respect to any requirement applicable to a cable operator, that it is commercially impracticable for the operator to comply with such requirement as a result of a change in conditions which is beyond the control of the operator and the nonoccurrence of which was a basic assumption on which the requirement was based. [47 USC 545]

Commission

Federal Communications Commission. [47 CFR 1.1803]

Committee of Principals (COP)

As specified by Executive Order 12472, a committee consisting of representatives from those federal departments, agencies or entities, designated by the President, which lease or own telecommunications facilities or services of significance to national security or emergency preparedness, and, to the extent permitted by law, other Executive entities which bear policy, regulatory or enforcement responsibilities of importance to national security or emergency preparedness telecommunications capabilities. [47 CFR Pt. 216, App.]

common carrier (carrier) [1]

Any person engaged as a common carrier for hire, in interstate or foreign communication by wire or radio or in interstate or foreign radio transmission of energy, except where reference is made to common carriers not subject to this Act; but a person engaged in radio broadcasting shall not, insofar as such person is so engaged, be deemed a common carrier. [47 USC 153, 47 CFR 32.9000]

common carrier (carrier) [2]

Any common carrier engaged in interstate communication by wire or radio as defined in section 3(h) of the Communications Act of 1934, as amended (the Act), and any common carrier engaged in intrastate communication by wire or radio, notwithstanding sections 2(b) and 221(b) of the Act. [47 USC 225, 47 CFR 64.601]

common channel network signaling

Channels between switching offices used to transmit signaling information independent of the subscribers' communication paths or transmission channels. [47 CFR Pt. 36, App.]

commonly owned carriers

Two or more carriers, one of which directly or indirectly owns more than 50 percent of the stock of the other carrier or carriers, or 50 percent or more of whose stock is owned directly or indirectly by the same person. [47 CFR 62.2]

communication by radio (see radio communication)

communication by wire (see wire communication)

Communications Act

The Federal Communications Act of 1934, as amended, 47 U.S.C. 151 *et seq.* [47 CFR 19.735-102]

communications assistant (CA)

A person who transliterates conversation from text to voice and from voice to text between two end users of TRS. CA supersedes the term "TDD operator." [47 CFR 64.601]

communications channel

In the Cellular Radiotelephone and Air-ground Radiotelephone Services, a channel used to carry subscriber communications. [47 CFR 22.99]

communication common carrier [1]

Any person engaged in rendering communication service for hire to the public. [47 CFR 21.2]

communications common carrier [2]

Any person (individual, partnership, association, joint-stock company, trust, corporation, or other entity) engaged as a common carrier for hire, in interstate or foreign communication by wire or radio or in interstate or foreign communication by wire or radio or in interstate or foreign radio transmission of energy, including such carriers as are described in subsection 2(b) (2) and (3) of the Communications Act of 1934, as amended, and, in addition, for purposes of Subpart H of this part, includes any individual, partnership, association, joint-stock company, trust, corporation, or other entity which owns or controls, directly or indirectly, or is under direct or indirect common control with, any such carrier. [47 CFR 25.103]

communications satellite corporation

(1) The corporation created pursuant to the provisions of Title III of the Communications Satellite Act of 1962. (2) The corporation shall be deemed to be a common carrier within the meaning of section 3(h) of the Communications Satellite Act of 1962. [47 CFR 25.103]

communication-satellite earth station complex

Includes transmitters, receivers, and communications antennas at the earth station site together with the interconnecting terrestrial facilities (cables, lines, or microwave facilities) and modulating and demodulating equipment necessary for processing of traffic received from the terrestrial distribution system(s) prior to transmission via satellite and of traffic received from the satellite prior to transfer of channels of communication to terrestrial distribution system(s). [47 CFR 25.103]

communication-satellite earth station complex functions

The communication-satellite earth station complex interconnects with terminal equipment of common carriers or authorized entities at the interface; accepts traffic from such entities at the interface, processes for transmission via satellite and performs the transmission function; receives traffic from a satellite or satellites, processes it in a form necessary to deliver channels of communication to terrestrial common carriers or such other authorized entities and delivers the processed traffic to such entities at the interface. [47 CFR 25.103]

community reception (in the broadcasting-satellite service)

The reception of emissions from a space station in the broadcasting-satellite service by receiving equipment, which in some cases may be complex and have antennae larger than those for individual reception, and intended for use (1) by a group of the general public at one location; or (2) through a distribution system covering a limited area. (RR) [47 CFR 2.1]

companion terminal equipment

Companion terminal equipment represents the terminal equipment that would be connected at the far end of a network facility and provides the range of operating conditions that the terminal equipment which is being registered would normally encounter. [47 CFR 68.3]

company or the company

When not otherwise indicated in the context, means the accounting entity. It includes such unincorporated entities which may be subject to the Communications Act of 1934, as amended. [47 CFR 32.9000]

comparable arrangement (see presubscription)

competing distributors

As used with respect to competing multichannel video programming distributors, means distributors whose actual or proposed service areas overlap. [47 CFR 76.1000]

complainant

A cable television system operator, a cable television system association, a utility, or an association of utilities who files a complaint. [47 CFR 1.1402]

complaint

A filing by a cable television system operator, a cable television system association, a utility, or an association of utilities alleging that a rate, term, or condition for a pole attachment is not just and reasonable. [47 CFR 1.1402]

complement (of cable)

A group of conductors of the same general type (*e.g.*, quadded, paired) within a single cable sheath. [47 CFR Pt. 36, App.]

complete complaint

A written statement that contains the complainant's name and address and describes the Commission's alleged discriminatory action in sufficient detail to inform the Commission of the nature and date of the alleged violation of section 504. It shall be signed by the complainant or by someone authorized to do so on his or her behalf. Complaints filed on behalf of classes or third parties shall describe or identify (by name, if possible) the alleged victims of discrimination. [47 CFR 1.1803]

complex

All groups of operator positions, wherever located, associated with the same call distribution and/or stored program control unit. [47 CFR Pt. 36, App.]

composite baseband signal

A signal which is composed of all program and other communications signals that frequency modulates the FM carrier. [47 CFR 73.310]

concentrating unit (TWX)

An arrangement of central office equipment wherein traffic over a number of TWX circuits is automatically concentrated onto a lesser number of circuits between the concentrating unit and its associated TWX switching office. [47 CFR Pt. 36, App.]

concentration equipment

Central office equipment whose function is to concentrate traffic from subscriber lines onto a lesser number of circuits between the remotely located concentration equipment and the serving central office concentration equipment. This concentration equipment is connected to the serving central office line equipment. [47 CFR Pt. 36, App.]

concurring carrier

A carrier (other than a connecting carrier) subject to the Act which concurs in and assents to schedules of rates and regulations filed on its behalf an issuing carrier or carriers. [47 CFR 61.3]

connecting carrier

(1) Any carrier engaged in interstate or foreign communication solely through physical connection with the facilities of another carrier not directly or indirectly controlling or controlled by, or under direct or indirect common control with such carrier [47 CFR 61.3], or (2) any carrier engaged in interstate or foreign communication solely through connection by radio, or by wire and radio, with facilities, located in an adjoining state or in Canada or Mexico (where they adjoin the state in which the carrier is doing business), or another carrier not directly or indirectly controlling or controlled by, or under

direct or indirect common control with such carrier, or (3) any carrier to which clause (1) or clause (2) would by applicable except for furnishing interstate mobile radio communication service or radio communication service to mobile stations on land vehicles in Canada or Mexico, except that Sections 201 through 205 of this Act, both inclusive, shall, except as otherwise provided therein, apply to carriers described in clauses (1), (2) and (3). [47 USC 153]

connection—minute

The product of (a) the number of messages and, (b) the average minutes of connection per message. [47 CFR Pt. 36, App.]

consent order

Is a formal decree accepting an agreement between a party to an adjudicatory hearing proceeding held to determine whether that party has violated statutes or Commission rules or policies and the appropriate operating Bureau, with regard to such party's future compliance with such statutes, rules or policies, and disposing of all issues on which the proceeding was designated for hearing. The order is issued by the officer designated to preside at the hearing or (if no officer has been designated) by the Chief Administrative Law Judge. [47 CFR 1.93]

consortium of small businesses owned by members of minority groups and/or women

A conglomerate organization formed as a joint venture between mutually-independent business firms, each of which individually satisfies the definitions in paragraphs (b) and (c) of this section. [47 CFR 24.720]

construction

As applied to public telecommunications facilities, acquisition (including acquisition by lease), installation, and modernization of public telecommunications facilities and planning and preparatory steps incidental to any such acquisition, installation, or modernization. [47 USC 397]

construction period

The period between the date of grant of an authorization and the date of required commencement of service. [47 CFR 22.99]

construction permit

That instrument of authorization required by the Act or the rules and regulations of the Commission made pursuant to this Act for the construction of a station or the installation of apparatus, for the transmission of energy, or communications, or signals by radio, by whatever name the instrument may be designated by the Commission. [47 USC 153]

consumer

A person initiating any interstate telephone call using operator services. [47 USC 226, 47 CFR 64.708]

consumer ISM equipment

A category of ISM Equipment used or intended to be used by the general public in a residential environment, notwithstanding use in other areas. Examples are domestic microwave ovens, jewelry cleaners for home use, ultrasonic humidifiers. [47 CFR 18.107]

consumer reporting agency

Has the meaning provided in the Federal Claims Collection Act, as amended, at 31 U.S.C. 3701(a)(3) or the Fair Credit Reporting Act, at 15 U.S.C. 168a(f). [47 CFR 1.1918]

contest [1]

Any contest broadcast by a radio station in connection with which any money or any other thing of value is offered as a prize or prizes to be paid or presented by the program sponsor or by any other person or persons, as announced in the course of the broadcast. [47 USC 508]

contest [2]

A scheme in which a prize is offered or awarded, based upon chance, diligence, knowledge or skill, to members of the public. [47 CFR 73.1225]

continuing financial interest

A right to receive revenue from the non-network broadcast or syndicated use of a television program by a network. [47 CFR 73.662]

continuity leads

Terminal equipment continuity leads at the network interface designated CY1 and CY2 which are connected to a strap in a series jack configuration for the purpose of determining whether the plug associated with the terminal equipment is connected to the interface jack. [47 CFR 68.3]

contract-based tariff

A tariff based on a service contract entered into between an interexchange carrier subject to §61.42 (a) through (c) or a nondominant carrier and a customer. [47 CFR 61.3]

control

Has the meaning that it has in 17 CFR §240.12b-2, the regulations promulgated by the Securities and Exchange Commission pursuant to the Securities Exchange Act of 1934 (15 USC 78a et seq.) Or any successor provision to such section. [47 USC 274]

control (including the terms "controlling", "controlled by," and "under common control with")

The possession directly or indirectly, or the power to direct or cause the direction of the management and policies of a company, whether such power is exercised through one or more intermediary companies, or alone, or in conjunction with, or pursuant to an agreement with, one or more other companies, and whether such power is established through a majority or minority ownership or voting of securities, common directors, officers, or stockholders, voting trusts, holding trusts, affiliated companies, contract, or any other direct or indirect means. [47 CFR 32.9000]

control channel

In the Cellular Radiotelephone Service and the Air-ground Radiotelephone Service, a channel used to transmit information necessary to establish or maintain communications. In the other Public Mobile Services, a channel that may be assigned to a control transmitter. [47 CFR 22.99]

control group

An entity, or a group of individuals or entities, that possesses *de jure* control and *de facto* control of an applicant or licensee, and as to which the applicant's or licensee's charters, articles of incorporation, bylaws, agreements and any other relevant documents (and amendments thereto) provide (1) that the entity and/or its members own unconditionally at least 50.1 percent of the total voting interests of a corporation; (2) that the entity and/or its members receive at least 50.1 percent of the annual distribution of any dividends paid on the voting stock of a corporation; (3) that, in the event of dissolution or liquidation of a corporation, the entity and/or its members are entitled to receive 100 percent of the value of each share of stock in its possession and a percentage of the retained earnings of the concern that is equivalent to the amount of equity held in the corporation; and (4) that the entity and/or its members have the right to receive dividends, profits and regular and liquidating distributions from the business in proportion to its interest in the total equity of the applicant or licensee. Note: Voting control does not always assure *de facto* control, such as, for example, when the voting stock of the control group is widely dispersed (see, *e.g.*, §24.720(e)(2)(iii)).[47 CFR 24.320, 24.720]

control operator
An amateur operator designated by the licensee of a station to be responsible for the transmissions from that station to assure compliance with the FCC Rules. [47 CFR 97.3]

control point [1]
A control point is an operating position at which an operator responsible for the operation of the transmitter is stationed and which is under the control and supervision of the licensee. [47 CFR 21.2]

control point [2]
A location where the operation of a public mobile station is supervised and controlled by the licensee of that station. [47 CFR 22.99]

control point [3]
Any place from which a transmitter's functions may be controlled. [47 CFR 90.7]

control point [4]
The location at which the control operator function is performed. [47 CFR 97.3]

control station [1]

A fixed station whose transmissions are used to control automatically the emissions or operations of another radio station at a specified location, or to transmit automatically to an alarm center telemetering information relative to the operation of such station. [47 CFR 21.2]

control station [2]

An operational fixed station the transmissions of which are used to control automatically the emissions or operation of another radio station at a specified location. [47 CFR 90.7]

control station [3]

An operational-fixed station the transmissions of which are used to control automatically the emissions or operations of a radio station in the mobile services at a specified location. [47 CFR 94.3]

control and telemetry transmissions

Signals transmitted on a multiplex subcarrier intended for any form of control and switching functions or for equipment status data and aural or visual alarms. [47 CFR 73.310]

control transmitter

A fixed transmitter in the public mobile services that transmits control signals to one or more base or fixed stations for the purpose of controlling the operation of the base or fixed stations, and/or transmits subscriber communications to one or more base or fixed stations that retransmit them to subscribers. [47 CFR 22.99]

CONV

International Telecommunications Convention, Malaga-Torremolinos, 1973. [47 CFR 2.1]

conventional radio system

A method of operation in which one or more radio frequency channels are assigned to mobile and base stations but are not employed as a trunked group. An "urban-conventional system" is one whose transmitter site is located within 24 km (15 miles) of the geographic center of any of the first 50 urbanized areas (ranked by population) of the United States. A "sub-urban-conventional system" is one whose transmitter site is located more than 24 km

(15 miles) from the geographic center of the first 50 urbanized areas. See Table 21, Rank of Urbanized Areas in the United States by Population, page 1-87, U.S. Census (1970); and Table 1 of §90.635. [47 CFR 90.7]

conversation—minute

The product of (a) the number of messages and, (b) the average minutes of conversation per message. [47 CFR Pt. 36, App.]

conversation-minute-kilometers

The product of (a) the number of messages, (b) the average minutes of conversation per message and (c) the average route kilometers of circuits involved. [47 CFR Pt. 36, App.]

coordinatable PCS device

PCS devices whose geographical are of operation is sufficiently controlled either by necessity of operation with a fixed infrastructure or by disabling mechanisms to allow adequate coordination of their locations relative to incumbent fixed microwave facilities. [47 CFR 15.303]

coordinated universal time (UTD)

Time scale, based on the second (SI), as defined and recommended by the CCIR, and maintained by the Bureau International de l'Heure (BIH). Note: For most practical purposes associated with the Radio Regulations, UTC is equivalent to mean solar time at the prime meridian (0 degrees longitude), formerly expressed in GMT. (RR) [47 CFR 2.1]

coordination area

The area associated with an earth station outside of which a terrestrial station sharing the same frequency band neither causes nor is subject to interfering emissions greater than a permissible level. (RR) [47 CFR 2.1]

coordination contour

The line enclosing the coordination area. (RR) [47 CFR 2.1]

coordination distance [1]

Distance on a given azimuth from an earth station beyond which a terrestrial causes nor is subject to interfering emissions greater than a permissible level. (RR) [47 CFR 2.1]

coordination distance [2]

For the purpose of this part, the expression "coordination distance" means the distance from an earth station, within which there is a possibility of the use of a given transmitting frequency at this earth station causing harmful interference to stations in the fixed or mobile service, sharing the same band, or of the use of a given frequency for reception at this earth station receiving harmful interference from such stations in the fixed or mobile service. [47 CFR 21.2]

coordination distance [3]

The distance from an earth station, within which there is a possibility of the use of a given transmitting frequency at this earth station causing harmful interference to stations in the fixed or mobile service sharing the same band, or of the use of a given frequency for reception at this earth station receiving harmful interference from such stations in the fixed or mobile service. [47 CFR 25.201]

cordless telephone system

A system consisting of two transceivers, one a base station that connects to the public switched telephone network and the other a mobile handset unit that communicates directly with the base station. Transmissions from the mobile unit are received by the base station and then placed on the public switched telephone network. Information received from the switched telephone network is transmitted by the base station to the mobile unit. [47 CFR 15.3]

corporate operations expenses

Includes Executive and Planning Expenses (Account 6710) and General and Administrative Expenses (Account 6720). [47 CFR 69.2]

corporation [1]

Any corporation, joint-stock company or association. [47 USC 153]

Corporation [2]

Corporation for Public Broadcasting authorized to be established in Subpart D. [47 USC 397]

Corporation [3]

The National Education Technology Funding Corporation described in subsection (a)(1)(A). [47 USC 708]

corrections

The remedy of errors in typing, spelling, or punctuation. [47 CFR 61.3]

cost [1]

Except as applied to telecommunications plants, franchises, and patent rights, means the amount of money actually paid (or the current money value of any consideration other than money exchanged) for property or services. See also Original Cost. [47 CFR 32.9000]

cost [2]

The cost of property owned by the Telephone Company whose property is to be apportioned among the operations. This term applies either to property costs recorded on the books of the company or property costs determined by other evaluation methods. [47 CFR Pt. 36, App.]

cost of removal

The cost of demolishing, dismantling, removing, tearing down, or otherwise disposing of telecommunications plant and recovering the salvage, including the cost of transportation and handling incident thereto. [47 CFR 32.9000]

CPI-U

Consumer Price Index for all Urban Consumers [47 CFR 1.1115]

CPST

Cable programming service tier [47 CFR 76.922]

CRAF

Civil Reserve Air Fleet [47 CFR 87.393]

creditor agency

The agency to which the debt is owed. [47 CFR 1.1901]

crime

Any act or omission that makes the offender subject to criminal punishment by law. [47 CFR 73.1225]

critical directional antenna

An AM broadcast directional antenna that is required, by the terms of a station authorization, to be operated with the relative currents and phases within the antenna elements at closer tolerances of deviation than those permitted under §73.62 and observed with a high precision monitor capable of measuring these parameters. [47 CFR 73.14]

critical hours

The two hour period immediately following local sunrise and the two hour period immediately preceding local sunset. [47 CFR 73.14]

cross-talk

An undesired signal occurring in one channel caused by an electrical signal in another channel. [47 CFR 73.310]

crystal

Quartz piezo-electric element. [47 CFR 95.669]

crystal controlled

Use of a crystal to establish the transmitted frequency. [47 CFR 95.669]

CSCE

Certificate of successful completion of an examination. [47 CFR 97.3]

CSTD

Cable system terminal device [47 CFR 15.3]

CTSs

Cell transmitter stations [47 CFR 95.803]

current billing

The combined amount of charges billed, excluding arrears. [47 CFR Pt. 36, App.]

customer

A user purchasing telecommunications service from a common carrier. [47 CFR 63.100]

customer dialed charge traffic

Traffic which is both (a) handled to completion through pulses generated by the customer and (b) for which either a message unit change, bulk charge or message toll charge is except for that traffic recorded by means of message registers. [47 CFR Pt. 36, App.]

customer operations expenses

Include Marketing and Services expenses in Accounts 6610 and 6620, respectively. [47 CFR 69.2]

customer premises equipment [1]

Equipment employed on the premises of person (other than a carrier) to originate, route, or terminate telecommunications. [47 USC 153]

customer premises equipment [2]

Items of telecommunications terminal equipment in Accounts 2310 referred to as CPE in §64.702 of the Federal Communication Commission's Rules adopted in the *Second Computer Inquiry* such as telephone instruments, data sets, dialers and other supplemental equipment, and PBX's which are provided by common carriers and located on customer premises and inventory included in these accounts to be used for such purposes. Excluded from this classification are similar items of equipment located on telephone company premises and used by the company in the normal course of business as well as over voltage protection equipment, customer premises wiring, coin operated public or pay telephones, multiplexing equipment to deliver multiple channels to the customer, mobile radio equipment and transmit earth stations. [47 CFR Pt. 36, App.]

customer premises wire

The segment of wiring from the customer's side of the protector to the customer premises equipment. [47 CFR Pt. 36, App.]

customer proprietary network information

(A) Information that relates to the quantity, technical configuration, type, destination, and amount of use of a telecommunications service subscribed to by any customer of a telecommunications carrier, and that is made available to the carrier by the customer solely by virtue of the carrier-customer relationship; and (B) information contained in the bills pertaining to telephone exchange service or telephone toll service received by a customer of a carrier; except that such term does not include subscriber list information. [47 USC 222]

C&WF

Categories of cable and wire facilities [47 CFR 36.152, 69.305]

D

DAEO
Designated agency ethics official [47 CFR 19.735-104]

data
Data configurations are those which use jacks incorporating components to limit signal power levels of data equipment, Data equipment with a maximum signal power output of -9 dBm may be connected to other than data configurations. See §68.308 of these rules. [47 CFR 68.502]

data processing equipment
Office equipment such as that using punched cards, punches tape, magnetic or other comparable storage media as an operating vehicle for recording and processing information. Includes machines for transcribing raw data into punched cards, etc., but does not include such items as key-operated, manually- or electrically-driven adding, calculating, bookkeeping or billing machines, typewriters or similar equipment. [47 CFR Pt. 36, App.]

date of commencement of the current license term
The date of commencement of the last term of license for which the licensee has been granted a license by the Commission. A separate license term shall not be deemed to have commenced as a result of continuing a license in effect under section 307(c) pending decision on an application for renewal of the license. [47 CFR 1.80]

day [1]
Any twenty-four hour period beginning 0100 g.m.t. and ending 0100 g.m.t. [47 CFR 73.701]

day [2]

(1) Where the word *day* is applied to the use of a specific frequency assignment or to a specific authorized transmitter power, its use means transmission on the frequency assignment or with the authorized transmitter power during that period of time included between one hour after local sunrise and one hour before local sunset: (2) Where the word *day* occurs in reference to watch requirements, or to equipment testing, its use means the calendar day, from midnight to midnight, local time. [47 CFR 80.5]

daytime

The period of time between local sunrise and local sunset. [47 CFR 73.14]

dB

Decibels [47 CFR 95.669]

DCW

Digital chart of the world [47 CFR 24.53]

dead spots

Small areas within a service area where the field strength is lower than the minimum level for reliable service. Service within dead spots is presumed. [47 CFR 22.99]

decision-making personnel

Any member, officer or employee of the Commission who is or may reasonably be expected to be involved in the decisional process in the proceeding. Unless otherwise specified, such persons usually include the Commissioners, their assistants, and other professional personnel of the Commission. Any person who has been made a party to a proceeding or who otherwise has been excluded from the decisional process shall not be treated as a decision-maker with the respect to that proceeding. Thus, any person designated as part of a separated trial staff shall not be considered a decision-making person in the designated proceeding. Unseparated Bureau or Office staff who may reasonably be expected to become involved in the decisional process of the proceeding shall be considered decision-making personnel. [47 CFR 1.1202]

dedicated signaling transport
Transport of out-of-band signaling information between an interexchange carrier or other person's common channel signaling network and a telephone company's signaling transport point on facilities dedicated to the use of a single customer. [47 CFR 69.2]

deep space
Space at distance from the earth equal to, or greater than, 2 x 10 to the sixth power kilometers. (RR) [47 CFR 2.1]

DEIS
Draft environmental impact statement [47 CFR 1.1305]

delinquent
A claim or debt which has not been paid by the date specified in the agency's written notification or applicable contractual agreement, unless other satisfactory payment arrangements have been made by that date, or, at any time thereafter, the debtor has failed to satisfy an obligation under a payment agreement with the agency. [47 CFR 1.1901]

delivered median field strength, or field strength
The field strength incident upon the zone or area of reception expressed in microvolts per meter, or decibels above one microvolt per meter, which is exceeded by the hourly median value for 50 percent of the days of the reference month. [47 CFR 73.701]

demarcation point [1]
The point of demarcation and/or interconnection between telephone company communications facilities and terminal equipment, protective apparatus or wiring at a subscriber's premises. Carrier-installed facilities at, or constituting, the demarcation point shall consist of wire or a jack conforming to subpart F of part 68 of the Commission's rules. [47 CFR 68.3]

demarcation point [2]
(1) For new and existing single unit installations, the demarcation point shall be a point at (or about) twelve inches outside of where the cable wire enters the subscriber's premises. (2) For new and existing multiple unit installations, the demarcation point shall be a point at (or about) twelve inches

outside of where the cable wire enters the subscriber's dwelling unit, but shall not include loop through or other similar series cable. [47 CFR 76.5]

DEMS

Digital electronic message service [47 CFR 21.6]

depreciation

The loss not restored by current maintenance, incurred in connection with the consumption or prospective retirement of telecommunications plant in the course of service from causes which are known to be in current operation, against which the company is not protected by insurance, and the effect of which can be forecast with a reasonable approach to accuracy. Among the causes to be given consideration are wear and tear, decay, action of the elements, inadequacy, obsolescence, changes in technology, changes in demand and requirements of public authorities. [47 CFR 32.9000]

designated authority

Any person, panel, or board which has been authorized by rule or order to exercise authority under section 5(d) of the Communications Act. [47 CFR 1.101]

designated community in a major television market

A community listed in §76.51. [47 CFR 76.5]

designated frequency

Each of the frequencies designated by the Commission for licenses for advanced television services. [47 USC 335]

developmental operation

A specially licensed operation for the purpose of testing concepts in the use of radio appropriate to the radio services governed by this part. [47 CFR 90.7]

dial switching equipment

Switching equipment actuated by electrical impulses generated by a dial or key pulsing arrangement. [47 CFR Pt. 36, App.]

dialing parity

A person that is not an affiliate of a local exchange carrier is able to provide telecommunications services in such a manner that customers have the ability to route automatically, without the use of any access code, their telecommunications to the telecommunications services providers of the customer's designation from among 2 or more telecommunications services providers (including such local exchange carrier). [47 USC 153]

DID

Direct inward dialing [47 CFR 68.312]

digital device

An unintentional radiator (device or system) that generates and uses timing signals or pulses at a rate in excess of 9,000 pulses (cycles) per second and uses digital techniques; inclusive of telephone equipment that uses digital techniques or any device or system that generates and uses radio frequency energy for the purpose of performing data processing functions, such as electronic computations, operations, transformations, recording, filing, sorting, storage, retrieval, or transfer. A radio frequency device that is specifically subject to an emanation requirement in any other FCC Rule part or an intentional radiator subject to subpart C of this part that contains a digital device is not subject to the standards for digital devices, provided the digital device is used only to enable operation of the radio frequency device and the digital device does not control additional functions or capabilities. [47 CFR 15.3]

digital electronic message service

A two-way domestic end-to-end fixed radio service utilizing digital termination systems for the exchange of digital information. This service may also make use of point-to-point microwave facilities, satellite facilities or other communications media to interconnect digital termination systems to comprise a network. [47 CFR 21.2]

digital milliwatt

A digital signal that is the coded re presentation of a 0 dBm, 1000 Hertz sine wave. [47 CFR 68.3]

digital modulation

The process by which some characteristic (frequency, phase, amplitude or combinations thereof) of a carrier frequency is varied in accordance with a digital signal, *e.g.,* one consisting of coded pulses or states. [47 CFR 21.2, 94.3]

digital selective calling (DSC)

A synchronous system developed by the International Radio Consultative Committee (CCIR), used to establish contact with a station or group of stations automatically by means of radio. The operational and technical characteristics of this system are contained in CCIR Recommendation 493. [47 CFR 80.5]

digital termination nodal station

A fixed point-to-multipoint radio station in a Digital Termination System providing two-way communication with Digital Termination User Stations. [47 CFR 21.2, 94.3]

digital termination system

A fixed point-to-multipoint radio system consisting of Digital Termination Nodal Stations and their associated Digital Termination User Stations. [47 CFR 21.2, 94.3]

digital termination user station

Any one of the fixed microwave radio stations located at users' premises, lying within the coverage area of a Digital Termination Modal Station, and providing two-way digital communications with the Digital Termination Nodal Station. [47 CFR 21.2, 94.3]

direct broadcast satellite service

A radiocommunication service in which signals transmitted or retransmitted by space stations are intended for direct reception by the general public. [47 CFR 100.3]

direct connection

Connection of terminal equipment to the telephone network by means other than acoustic and/or inductive coupling. [47 CFR 68.3]

direct costs

Those expenditures which the Commission actually incurs in searching for and duplicating (and in case of commercial requesters, reviewing) documents to respond to a FOIA request. Direct costs include the salary of the employee performing the work the employee performing the work (the basic rate of pay for the employee plus 16 percent of that rate to cover benefits), and the cost of operating duplicating machinery. Not included in direct costs are overhead expenses, such as costs of space, and heating or lighting the facility in which the records are stored. [47 CFR 0.466]

direct expense

Expenses that are attributable to a particular category or categories of tangible investment described in subpart D of this part and includes (1) Plant Specific Operations Expenses in Accounts 6110, 6120, 6210, 6220, 6230, 6310, 6410; and (2) Plant Nonspecific Operations Expenses in Accounts 6510, 6530, 6540 and 6560. [47 CFR 69.2]

direct reception [1]

In the broadcasting-satellite service, shall encompass both individual reception and community reception. (RR) [47 CFR 2.1]

direct reception [2]

Shall encompass both individual reception and community reception. [47 CFR 100.3]

direct sequence systems

A direct sequence system is a spread spectrum system in which the incoming information is usually digitized, if it is not already in a binary format, and modulo 2 added to a higher speed code sequence. The combined information and code are then used to modulate a RF carrier. Since the high speed code sequence dominates the modulating function, it is the direct cause of the wide spreading of the transmitted signal. [47 CFR 2.1]

direct-to-home satellite services

The distribution or broadcasting of programming or services by satellite directly to the subscriber's premises without the use of ground receiving or distribution equipment, except at the subscriber's premises or in the uplink process to the satellite. [47 USC 303]

direct-trunked transport

Transport on circuits dedicated to the use of a single interexchange carrier or other person, without switching at the tandem, (1) between the serving wire center and the end office, or (2) between two customer-designated telephone company offices. [47 CFR 69.2]

direction finder (radio compass)

Apparatus capable of receiving radio signals and taking bearings on these signals from which the true bearing and direction of the point of origin may be determined. [47 CFR 80.5]

director (see officer)

disability

Has the meaning given to it by Section 3(2)(A) of the Americans with Disabilities Act of 1990 (42 USC 12102(2)(A)). [47 USC 255]

discontinuance of service ("reduction" or "impairment of service").

Includes, but is not limited to the following:

(1) The closure by a carrier of a telephone exchange rendering interstate or foreign telephone toll service, a public toll station serving a community or part of a community, or a public coast station as defined in §80.5 of this chapter;

(2) The reduction in hours of service by a carrier at a telephone exchange rendering interstate or foreign telephone toll service, at any public toll station (except at a toll station at which the availability of service to the public during any specific hours is subject to the control of the agent or other persons controlling the premises on which such office or toll station is located and is not subject to the control of such carrier), or at a public coast station; the term *reduction in hours of service* does not include a shift in hours which does not result in any reduction in the number of hours of service.

(3) [Reserved]

(4) The dismantling or removal for service of any trunk line by a carrier which has the effect of impairing the adequacy or quality of service rendered to any community or part of a community;

(5) The severance by a carrier of physical connection with another carrier (including connecting carriers as defined in section 3(u) of the Communications Act of 1934, as amended) or the termination or suspension of the interchange of traffic with such other carrier. [47 CFR 63.60]

dispatch point

Any place from which radio messages can be originated under the supervision of a control point. [47 CFR 90.7]

dispatch service

A radiotelephone service comprising communications between a dispatcher and one or more mobile units. These communications normally do not exceed one minute in duration and are transmitted directly through a base station, without passing through mobile telephone switching facilities. [47 CFR 22.99]

display disable

To turn off the display of captions or text (and accompanying background) at the receiver, rather than through codes transmitted on line 21 which unconditionally erase the display. The receiver may disable the display because the user selects and alternate mode, *e.g.*, TV Mode, or because no valid line 21 data is present. [47 CFR 15.119]

display enable

To allow the display of captions or text when they are transmitted on line 21 and received as valid data. For display to be enabled, the user must have selected Caption Mode or Text Mode, and valid data for the selected mode must be present on line 21. [47 CFR 15.119]

displayable character

Any letter, number or symbol which is defined for on-screen display, plus the 20th space.
[47 CFR 15.119]

disposable pay

That part of current basic pay, special pay, incentive pay, retired pay, retainer pay, or in the case of an employee not entitled to basic pay, other authorized pay remaining after the deduction of any amount required by law to be withheld. Agencies must exclude deductions described in 5 CFR 581.105 (b) through (f) to determine disposable pay subject to salary offset. [47 CFR 1.1901]

distress signal

The distress signal is an internationally recognized radiotelegraph or radio telephone transmission which indicates that a ship, aircraft, or other vehicle is threatened by grave and imminent danger and requests immediate assistance. (1) In radiotelegraphy, the international distress signal consists of the group "three dots, three dashes, three dots," transmitted as a single signal in which the dashes are emphasized so as to be distinguished clearly from the dots. (2) In radiotelephony, the international distress signal consists of the enunciation of the word "Mayday," pronounced as the French expression "m'aider." In case of distress, transmission of this particular signal is intended to ensure recognition of radiotelephone distress call by stations of any nationality. [47 CFR 80.5]

distress traffic

All messages relative to the immediate assistance required by a ship, aircraft, or other vehicle in distress. [47 CFR 80.5]

DME

Distance measuring equipment [47 CFR 87.475]

DMO

Practice/demo warning [47 CFR 11.31]

DOJ

Department of Justice [47 CFR 1.1902]

domestic fixed public service

A fixed service, the stations of which are open to public correspondence, for radiocommunications originating and terminating solely at points all of which lie within (a) the State of Alaska, or (b) the State of Hawaii, or (c) the contiguous 48 States and the District of Columbia, or (d) a single possession of the United States. Generally, in cases where service is afforded on frequencies above 72 MHZ, radiocommunications between the contiguous 48 States (including the District of Columbia) and Canada or Mexico, or radiocommunications between the State of Alaska and Canada, are deemed to be in the domestic fixed public service. [47 CFR 21.2]

domestic public radio services
The land mobile and domestic fixed public services [are] the stations which are open to public correspondence. [47 CFR 21.2]

dominant carrier
A carrier found by the Commission to have market power (*i.e.*, power to control prices). [47 CFR 61.3]

DPU
Direct pickup [47 CFR 15.118]

drop point
A term used in the point-to-point microwave radio service to designate a terminal point where service is rendered to a subscriber. [47 CFR 21.2]

DSA board
A local dial office switchboard at which are handled assistance calls, intercepted calls and calls from miscellaneous lines and trunks. It may also be employed for handling certain toll calls. [47 CFR Pt. 36, App.]

DSB board
A switchboard of a dial system for completing incoming calls received from manual offices. [47 CFR Pt. 36, App.]

DSC
Digital selective calling [47 CFR 80.5]

DTMF
Dual tone multifrequency [47 CFR 68.308]

duplex operation
Operating method in which transmission is possible simultaneously in both directions of a telecommunication channel. (RR) [47 CFR 2.1]

E

EA

Economic area [47 CFR 26.102]

EAH

Effective antenna height [47 CFR 90.279]

EAJA

Equal Access to Justice Act [47 CFR 1.1501]

EAN

Emergency action notification [47 CFR 11.31]

earth exploration-satellite service

A radiocommunication service between earth stations and one or more space stations, which may include links between space stations in which (1) information relating to the characteristics of the earth and its natural phenomena is obtained from active sensors or passive sensors on earth satellites; (2) similar information is collected from air-borne or earth-based platforms; (3) such information may be distributed to earth stations within the system concerned; (4) platform interrogation may be included. Note: This service may also include feeder links necessary for its operation. (RR) [47 CFR 2.1]

earth station [1]

A station located either on the earth's surface or within the major portion of the earth's atmosphere intended for communication (a) with one or more space stations; or (b) with one or more stations of the same kind by means of one or more reflecting satellites or other objects in space. [47 CFR 2.1, 25.201]

earth station [2]

A station in the space service located either on the earth's surface, including on board a ship, or on board an aircraft. [47 CFR 21.2]

earth station [3]

An amateur station located on, or within 50 km of, the earth's surface intended for communications with space stations or with other earth stations by means of one or more other objects in space. [47 CFR 97.3]

EAs

Environmental assessments [47 CFR 1.1307]

EAS

Emergency alert system [47 CFR 0.183], extended area services [47 CFR 36.392]

EAT

Emergency action termination [47 CFR 11.31]

educational institution

A preschool, a public or private elementary or secondary school, an institution or graduate higher education, an institution of professional education and an institution of vocational education, which operates a program or programs of scholarly research. [47 CFR 0.466]

educational television programming for children

Any television program which is directed to an audience of children who are 16 years of age or younger and which is designed for the intellectual development of those children, except that such term does not include any television program which is directed to a general audience but which might also be viewed by a significant number of children. [47 USC 394]

EEOC

Equal Employment Opportunity Commission [47 CFR 22.321]

effective competition

(A) Fewer than 30 percent of the households in the franchise area subscribe to the cable service of a cable system;

(B) the franchise area is (i) served by at least two unaffiliated multichannel video programming distributors each of which offers comparable video programming to at least 50 percent of the households in the franchise area; and (ii) the number of households subscribing to programming services offered by multichannel video programming distributors other than the largest multichannel video programming distributor exceeds 15 percent of the households in the franchise area;

(C) a multichannel video programming distributor operated by the franchising authority for that franchise area offers video programming to at least 50 percent of the households in that franchise area; or

(D) a local exchange carrier or its affiliate (or any multichannel video programming distributor using the facilities of such carrier or its affiliate) offers video programming services directly to subscribers by any means (other than direct-to-home satellite services) in the franchise area of an unaffiliated cable operator which is providing cable service in that franchise area, but only if the video programming services so offered in that area are comparable to the video programming services provided by the unaffiliated cable operator in that area. [47 USC 543]

effective field (effective field strength)

The root-mean-square (RMS) value of the inverse distance fields at a distance of 1 kilometer from the antenna in all directions in the horizontal plane. The term "field strength" is synonymous with the term "field intensity" as contained elsewhere in the Part. [47 CFR 73.14]

effective isotropically radiated power (e.i.r.p.)

The product of the power supplied to the antenna and the antenna gain in a given direction relative to an isotropic antenna. [47 CFR 24.5)

effective radiated power [1]

The product of the antenna power input and the antenna power gain. This product should be expressed in watts. (If specified for a particular direction, effective radiated power is based on the antenna power gain in that direction only.) [47 CFR 21.2]

effective radiated power (ERP) [2]

The effective radiated power of a transmitter (with antenna, transmission line, duplexers, etc.) is the power that would be necessary at the input terminals of a reference half-wave dipole antenna in order to produce the same maximum field intensity. ERP is usually calculated by multiplying the measured transmitter output power by the specified antenna system gain, relative to a half-wave dipole, in the direction of interest. [47 CFR 22.99]

effective radiated power [3]

The product of the antenna power (transmitter output power less transmission line loss) times: (1) the antenna power gain, or (2) the antenna field gain squared. Where circular or elliptical polarization is employed, the term effective radiated power is applied separately to the horizontal and vertical components of radiation. For allocation purposes, the effective radiated power authorized is the horizontally polarized component of radiation only. [47 CFR 73.310]

effective radiated power [4]

The product of the antenna input power and the antenna power gain. This product should be expressed in kW and in dB above 1 kW (dBk). (If specified for a particular direction, effective radiated power is based on the antenna power gain in that direction only [47 CFR 21.2]. The licensed effective radiated power is based on the maximum antenna power gain. When a station is authorized to use a directional antenna or an antenna beam tilt, the direction of the maximum effective radiated power will be specified.) Where circular or elliptical polarization is employed, the term effective radiated power is applied separately to the horizontally and vertically polarized components of radiation. For assignment purposes, only the effective radiated power authorized for the horizontally polarized component will be considered. [47 CFR 73.681]

effective radiated power (ERP) [5]

The power supplied to an antenna multiplied by the relative gain of the antenna in a given direction. [47 CFR 90.7]

effective radiated power (in a given direction)

The product of the power supplied to the antenna and its gain relative to a half-wave dipole in a given direction. (RR) [47 CFR 2.1, 24.5]

EHF

Extremely high frequency [47 CFR 97.3]

EIA
Electronics Industries Association [47 CFR 24.237]

EIC
Engineer in charge [47 CFR 18.115]

EIRP (or e.i.r.p.)
Equivalent isotropically radiated power [47 CFR 21.2]

EIS
Environmental impact statement [47 CFR 1.1305]

election
(1) A general, special, primary, or runoff election, (2) a convention or caucus of a political party held to nominate a candidate, (3) a primary election held for the selection of delegates to a national nominating convention of a political party, and (4) a primary election held for the expression of a preference for the nomination of persons for election to the office of President. [47 CFR 64.803]

elective calling
A means of calling in which signals are transmitted in accordance with a prearranged code to operate a particular automatic attention device at the station whose attention is sought. [47 CFR 80.5]

electronic publishing
The dissemination, provision, publication, or sale to an unaffiliated entity or person, of any one or more of the following: news (including sports); entertainment (other than interactive games); business, financial, legal, consumer, or credit materials; editorials, columns, or features; advertising; photos or images; archival or research material; legal notices or public records; scientific, educational, instructional, technical, professional, trade, or other literary materials; or other like or similar information.
The term shall not include the following services:
(A) information access, as that term is defined by the AT&T Consent Decree.
(B) The transmission of information as a common carrier.
(C) The transmission of information as part of a gateway to an information service that does not involve the generation or alteration of the

content of information, including data transmission, address translation, protocol conversion, billing management, introductory information content, and navigational systems that enable users to access electronic publishing services, which do not affect the presentation of such electronic publishing services to users.

(D) Voice storage and retrieval services, including voice messaging and electronic mail services.

(E) Data processing or transaction processing services that do not involve the generation or alteration of the content of information.

(F) Electronic billing or advertising of a Bell operating company's regulated telecommunications services.

(G) Language translation or data format conversion.

(H) The provision of information necessary for the management, control, or operation of a telephone company telecommunications system.

(I) The provision of directory assistance that provides names, addresses, and telephone numbers and does not include advertising.

(J) Caller identification services.

(K) Repair and provisioning databases and credit card and billing validation for telephone company operations.

(L) 911-E and other emergency assistance databases.

(M) Any other network service of a type that is like or similar to these network services and that does not involve the generation or alteration of the content of information.

(N) Any upgrades to these network services that do not involve the generation or alteration of the content of information.

(O) Video programming or full motion video entertainment on demand. [47 USC 274]

electronic publishing joint venture
A joint venture owned by a Bell operating company or affiliate that engages in the provision of electronic publishing which is disseminated by means of such Bell operating company's or any of its affiliates' basic telephone service. [47 USC 274]

element

In a pop-on or paint-on style caption, each contiguous area of cells containing displayable characters and nontransparent spaces between those characters. A single caption may have multiple elements. An element is not necessarily a perfect rectangle, but may include rows of differing widths. [47 CFR 15.119]

elementary and secondary schools

Elementary schools and secondary schools, as defined in paragraphs (14) and (25), respectively, of Section 14101 of the Elementary and Secondary Education Act of 1965 (20 USC 8801). [47 USC 254]

eligible small business

Business enterprises engaged in the telecommunications industry that have $50,000,000 or less in annual revenues, on average over the past 3 years prior to submitting the application under this section. [47 USC 614]

ELT

Emission limitations [47 CFR 87.139]

EME

Electromagnetic emissions [47 CFR 2.1511]

E&M leads

Terminal equipment leads at the interface, other than telephone connections and auxiliary leads, which are to be connected to channel equipment solely for the purpose of transferring supervisory signals conventionally known as Types I and II E&M and schematically shown in Figures 68.3(e)(i) and 68.3(a)(ii). [47 CFR 68.3]

Emergency Alert System (EAS)

The EAS is composed of broadcast networks; cable networks and program suppliers; AM, FM and TV broadcast stations; Low Power TV (LPTV) stations; subject cable systems; and other entitles and industries operating on an organized basis during emergencies at the national, state, or local levels. [47 CFR 76.5]

emergency discontinuance of service ("emergency reduction" or "impairment of service")

Any discontinuance, reduction, or impairment of the service of a carrier occasioned by conditions beyond the control of such carrier where the original service is not restored or comparable service is not established within a reasonable time. For the purpose of this part, a reasonable time shall be deemed to be a period not in excess of the following: 10 days in the case of discontinuance, reduction, or impairment of service at telegraph offices operated directly by the carrier; 15 days in the case of jointly operated or agency telegraph offices; 10 days in the case of public coast stations; and 60 days in all other cases. [47 CFR 63.60]

emergency locator transmitter (ELT)

A transmitter of an aircraft or a survival craft actuated manually or automatically that is used as an altering and locating aid for survival purposes. [47 CFR 87.5]

emergency locator transmitter (ELT) test station

A land station used for testing ELTs or for training in the use of ELTs. [47 CFR 87.5]

emergency position indicating radio beacon station [1]

A station in the mobile service the emissions of which are intended to facilitate search and rescue operations. (RR) [47 CFR 2.1]

emergency position indicating radio beacon (EPIRB) station [2]

A station in the maritime mobile service the emissions of which are intended to facilitate search and rescue operations. [47 CFR 80.5]

emergency purposes

Calls made necessary in any situation affecting the health and safety of consumers. [47 CFR 64.1200]

emergency reduction of service (see emergency discontinuance of service)

emergency service

Service for which there is an immediate need occasioned by conditions unforeseen by, and beyond the control of, the carrier. [47 CFR 63.04]

emerging network

An entity not meeting the definition of a "television network," as set forth in paragraph (f) of this section, on June 5, 1993, but which subsequently meets this definition. [47 CFR 73.662]

emission [1]

Radiation produced, or the production of radiation, by a radio transmitting station. Note: For example, the energy radiated by the local oscillator of a radio receiver would not be an emission but a radiation. (RR) [47 CFR 2.1]

emission [2]

The electromagnetic energy radiated from an antenna. [47 CFR 22.99]

emission bandwidth

For purposes of this subpart the emission bandwidth shall be determined by measuring the width of the signal between two points, one below the carrier center frequency and one above the carrier center frequency, that are 26 dB down relative to the maximum level of the modulated carrier. Compliance with the emissions limits is based on the use of measurement instrumentation employing a peak detector function with an instrument resolutions bandwidth approximately equal to 1.0 percent of the emission bandwidth of the device under measurement. [47 CFR 15.303]

emission designator

An internationally accepted symbol for describing an emission in terms of its bandwidth and the characteristics of its modulation, if any. See §2.201 of this chapter for details. [47 CFR 22.99]

emission mask

The design limits imposed, as a condition or type acceptance, on the mean power of emissions as a function of frequency both within the authorized bandwidth and in the adjacent spectrum. [47 CFR 22.99]

employee [1]

A current employee of the Commission or of another agency, including a current member of the Armed Forces or a Reserve of the Armed Forces (Reserve). [47 CFR 1.1901]

employee [2]

Includes furloughed, pensioned, and superannuated employees. [47 CFR 4.1]

employee [3]

An officer or employee of the Commission including the Commissioners, but does not include a special government employee or member of the uniformed services. [47 CFR 19.735-102]

encoded analog content

The analog signal contained in coded form within a digital signal. [47 USC 68.3]

encrypt

When used with respect to satellite cable programming, means to transmit such programming in a form whereby the aural and visual characteristics (or both) are modified or altered for the purpose of preventing the unauthorized receipt of such programming by persons without authorized equipment which is designed to eliminate the effects of such modification or alteration. [47 USC 605]

end office

The telephone company office from which the end user receives exchange service. [47 CFR 69.2]

end user

Any customer of an interstate or foreign telecommunications service that is not a carrier except that a carrier other than a telephone company shall be deemed to be an "end user" when such carrier uses a telecommunications service for administrative purposes and a person or entity that offers telecommunications services exclusively as a reseller shall be deemed to be an "end user" if all resale transmissions offered by such reseller originate on the premises of such reseller. [47 CFR 69.2]

entertainment programs

Include series, made-for-television movies, mini-series, and entertainment specials, and do not include sports, public affairs, or news programs. [47 CFR 73.662]

entity [1]

Any organization, and includes corporations, partnerships, sole proprietorships, associations, and joint ventures. [47 USC 274]

entity [2]

A legal enterprise (common carrier) engaged in interstate communications within the meaning of the Communications Act of 1934, as amended. [47 CFR 32.9000]

entrance facilities

Transport from the interexchange carrier or other person's point of demarcation to the serving wire center. [47 CFR 69.2]

entry switch

The telephone company switch in which a transport line or trunk terminates. [47 CFR 69.2]

environmental communications

Broadcasts of information about the environmental conditions in which vessels operate, *i.e.*, weather, sea conditions, time signals adequate for practical navigation, notices to mariners, and hazards to navigation. [47 CFR 80.5]

EOM

End of message [47 CFR 11.31]

EPIRB

Emergency position indicating radiobeacon [47 CFR 80.5]

equal access

The meaning given that term in Appendix B of the Modification of Final Judgment entered August 24, 1982, in United States v. Western Electric, Civil Action No. 82-0192 (United States District Court, District of Columbia), as amended by the Court in its orders issued prior to the enactment of this section. [47 USC 226, 47 CFR 64.708]

equal access code

An access code that allows the public to obtain an equal access connection to the carrier associated with that code. [47 CFR 64.708]

equal access costs

Include only initial incremental presubscription costs and initial incremental expenditures for hardware and software related directly to the provision of equal access which would not be required to upgrade the switching capabilities of the office involved absent the provisions of equal access. [47 CFR Pt. 36, App.]

equal access investment and equal access expenses

Equal access investment and expenses as defined for purposes of the part 36 separations rules. [47 CFR 69.2]

equipment performance measurements

The measurements performed to determine the overall performance characteristics of a broadcast transmission system from point of program origination at main studio to sampling of signal as radiated. (See §73.1590.) [47 CFR 73.14]

equivalent gauge

A standard cross section of cable conductors for use in equating the metallic content of cable conductors of all gauge to a common base. [47 CFR Pt. 36, App.]

equivalent isotropically radiated power (e.i.r.p.) [1]

The product of the power supplied to the antenna and the antenna gain in a given direction relative to an isotropic antenna. (RR) [47 CFR 2.1, 26.4]

equivalent isotropically radiated power (EIRP) [2]

The product of the power supplied to the transmitting antenna and the antenna gain in a given direction relative to an isotropic antenna radiator. This product may be expressed in watts or dB above 1 watt (dBW). [47 CFR 21.2, 73.310, 74.901]

equivalent isotropically radiated power (EIRP) [3]

The equivalent isotropically radiated power of a transmitter (with antenna, transmission line, duplexers, etc.) is the power that would be necessary at the input terminals of a reference isotropic radiator in order to produce the same maximum field intensity. An isotropic radiator is a theoretical lossless point source of radiation with unity gain in all directions. EIRP is usually calculated by multiplying the measured transmitter output power by the specified antenna system gain, relative to an isotropic radiator, in the direction of interest. [47 CFR 22.99]

equivalent isotropically radiated power (EIRP) [4]

The product of the power supplied to the antenna and the antenna gain in a given direction relative to an isotropic antenna. For purpose of this part, EIRP is expressed in decibels referenced to 1 milliwatt (dBm) in the direction of the main beam. [47 CFR 94.3]

equivalent kilometers of 104 wire

The basic units employed in the allocation of pole line costs for determining the relative use made of poles by aerial cables and by aerial wire conductors of various sizes. This unit reflects the relative loads of such cable and wire carried on poles. [47 CFR Pt. 36, App.]

equivalent monopole radiated power (e.m.r.p.) (in a given direction)

The product of the power supplied to the antenna and its gain relative to a short vertical antenna in a given direction. (RR) [47 CFR 2.1]

equivalent pair kilometers

The product of sheath kilometers and the number of equivalent gauge pairs of conductors in a cable. [47 CFR Pt. 36, App.]

equivalent power

The power of the analog signal at the output of a zero level decoder, obtained when a digital signal is the input to the decoder. [47 CFR 68.3]

equivalent satellite link noise temperature

The noise temperature referred to the output of the receiving antenna of the earth station corresponding to the radio-frequency noise power which produces the total observed noise at the output of the satellite link excluding the noise due to interference coming from satellite links using other satellites and form terrestrial systems. (RR) [47 CFR 2.1]

equivalent sheath kilometers

The product of (a) the length of a section of cable in kilometers (sheath kilometers) and (b) the ratio of the metallic content applicable to a particular group of conductors in the cable (*e.g.*, conductors assigned to a category) to the metallic content of all conductors in the cable. [47 CFR Pt. 36, App.]

erase display

In Caption Mode, to clear the screen of all characters (and accompanying background) in response to codes transmitted on line 21. (The caption service provider can accomplish the erasure either by sending an Erase Displayed Memory command or by sending an Erase Non-Displayed Memory command followed by an End of Caption command, effectively making a blank caption "appear".) Display can also be erased by the receiver when the caption memory erasure conditions are met, such as the user changing TV channels. [47 CFR 15.119]

ERP

Effective radiated power [47 CFR 22.99]

ESN

Electronic serial number [47 CFR 22.919]

ESPN

Entertainment and Sports Programming Network [47 CFR 11.43]

essential telephones

Only coin-operated telephones, telephones provided for emergency use, and other telephones frequently needed for use by persons using such hearing aids. [47 USC 605, 47 CFR 68.3]

established business relationship

A prior or existing relationship formed by a voluntary two-way communication between a person or entity and a residential subscriber with or without an exchange of consideration, on the basis of an inquiry, application, purchase or transaction by the residential subscriber regarding products or services offered by such person or entity, which relationship has not been previously terminated by either party. [47 CFR 64.1200]

ETS

Emerging technologies services [47 CFR 22.602]

EUCL

End user common line [47 CFR 61.47]

EVI

Evacuation immediate [47 CFR 11.31]

exceptions

The document consolidating the exceptions and supporting brief. The brief shall contain (i) a table of contents, (ii) a table of citations, (iii) a concise statement of the cast, (iv) a statement of the questions of law presented, and (v) the argument, presenting clearly the points of fact and law relied upon in support of the position taken on each question, with specific reference to the record and all legal or other materials relied on. [47 CFR 1.276]

exchange

A unit of a communication company or companies for the administration of communication service in a specified area, which usually embraces a city, town, or village and its environs, and consisting of one or more central offices, together with the associated plant, used in furnishing communication service in that area. [47 CFR 21.2]

exchange access

The offering of access to telephone exchange services or facilities for the purpose of the origination or termination of telephone toll services. [47 USC 153]

exchange area

The geographic area included within the boundaries of an exchange. [47 CFR 21.2]

exchange transmission plant

This is a combination of (a) exchange cable and wire facilities; (b) exchange central office circuit equipment, including associated land and building; and (c) information origination/termination equipment which forms a complete channel. [47 CFR Pt. 36, App.]

Executive Order

Executive Order 11222 of May 8, 1965. [47 CFR 19.735-102]

exempt telecommunications company

Any person determined by the Federal Communications Commission to be engaged directly or indirectly, wherever located, through one or more affiliates (as defined in section 2(a)(11)(B)), and exclusively in the business of providing (A) telecommunications services; (B) information services; (C) other services or products subject to the jurisdiction of the Federal Communications Commission; or (D) products or services that are related or incidental to the provision of a product or service described in subparagraph (A), (B), or (C). No person shall be deemed to be an exempt telecommunications company under this section unless such person has applied to the Federal Communications Commission for a determination under this paragraph. A person applying in good faith for such a determination shall be deemed an exempt telecommunications company under this section, with all of the exemptions provided by this section, until the Federal Communications Commission makes such determination. The Federal Communications Commission shall make such determination within 60 days of its receipt of any such application filed after the enactment of this section and shall notify the Commission whenever a determination is made under this paragraph that any person is an exempt telecommunications company. Not later than 12 months after the date of enactment of this section, the Federal Communications Commission shall promulgate rules implementing the provisions of this paragraph which shall be applicable to applications filed under this paragraph after the effective date of such rules. [47 USC 103]

ex parte presentation

Any presentation made to decision-making personnel but, in restricted proceedings, any presentation to or from decision-making personnel, which (1) if written, is not served on the parties to the proceeding, or (2) if oral, is made without advance notice to the parties to the proceedings and without opportunity for them to be present. Comments and reply comments (including informal comments) filed prior to the expiration of the reply comment period, or, if the matter is on reconsideration, the reconsideration reply comment period, in informal rulemaking proceedings pursuant to §§1.415 and 1.419, but not in channel allotment rulemaking proceedings pursuant to §1.420, are not considered ex parte presentations even if they are not served on other parties. [47 CFR 1.1202]

expendable launch vehicle (ELV)
A booster rocket that can be used only once to launch a payload, such as a missile or space vehicle. [47 CFR 87.5]

expense component
The total expenses and income charges for an annual period that are attributable to a particular element or category. [47 CFR 69.2]

expenses
Include allowable expenses in the Uniform System of Accounts, part 32, apportioned to interstate or international services pursuant to the *Separations Manual* and allowable income charges apportioned to interstate and international services pursuant to the *Separations Manual*. [47 CFR 69.2]

experimental broadcast station
A station licensed for experimental or developmental transmission of radio telephony, television, facsimile, or other types of telecommunication services intended for reception and use by the general public. [47 CFR 74.101]

experimental period
The time between 12 midnight local time and local sunrise, used by AM stations for tests, maintenance and experimentation. [47 CFR 73.14]

experimental radio service
A service in which radio waves are employed for purposes of experimentation in the radio art or for purposes of providing essential communications for research projects which could not be conducted without the benefit of such communications. [47 CFR 5.4]

experimental station
A station utilizing radio waves in experiments with a view to the development of science or technique. Note: This definition does not include amateur stations. (RR) [47 CFR 2.1, 5.4]

extension
In the Cellular Radio-Telephone Service, an area within the service area boundary of a cellular system, but outside of the market boundary. See §§22.911(c) and 22.912. [47 CFR 22.99]

external RF power amplifier

A device capable of increasing power output when used in conjunction with, but not an integral part of, a transmitter. [47 CFR 97.3]

external RF power amplifier kit

A number of electronic parts, which, when assembled, is an external RF power amplifier, even if additional parts are required to complete assembly. [47 CFR 97.3]

F

FAA

Federal Aviation Administration [47 CFR 1.61]

facility

All or any portion of buildings, structures, equipment, roads, walks, parking lots, rolling stock or other conveyances, or other real or personal property. [47 CFR 1.1803]

facsimile [1]

A form of telegraphy for the transmission of fixed images, with or without half-tones, with a view to their reproduction in a permanent form. Note: In this definition the term telegraphy has the same general meaning as defined in the Convention. (RR) [47 CFR 2.1]

facsimile [2]

A system of telecommunication for the transmission of fixed images with a view to their reception in a permanent form. [47 CFR 21.2]

facsimile service

Transmission of still images from one place to another by means of radio. [47 CFR 22.99]

families

The wives, husbands, minor children, and other dependents of the officers, employees, or agents permitted to receive and use franks, but no other person. [47 CFR 41.1]

FAR

Federal acquisition regulations [47 CFR 1.1902]

FASB 51

Financial Accounting Standards Board Standard 51 [47 CFR 76.922]

FCC

Federal Communications Commission [47 CFR 0.401]

FCCS

The Federal Claims Collection Standards jointly published by the Justice Department and the General Accounting Office at 4 CFR parts 101-105. [47 CFR 1.1901]

federal agency

Any agency of the United States, including the Commission. [47 USC 522]

federal office

The office of President or Vice President of the Untied States: or of Senator or Representative in, or Delegate or Resident Commissioner to, the Congress of the United States. [47 CFR 64.803]

fee (see tax)

feeder link

A radio link from an earth station at a given location to a space station, or vice versa, conveying information for a space radiocommunication service other than for the fixed-satellite service. The given location may be at a specified fixed point, or at any fixed point within specified areas. (RR) [47 CFR 2.1]

FEIS

Final environmental impact statement [47 CFR 1.1305]

FFA

Flash flood watch [47 CFR 11.31]

FFS

Flash flood statement [47 CFR 11.31]

FFW

Flash flood warning [47 CFR 11.31]

field

Scanning through the picture area once in the chosen scanning pattern. In the line interlaced scanning pattern of two to one, the scanning of the alternate lines of the picture area once. [47 CFR 73.681]

field disturbance sensor

A device that establishes a radio frequency field in its vicinity and detects changes in that field resulting from the movement of persons or objects within its range. [47 CFR 15.3]

field strength

The electric field strength in the horizontal plane. [47 CFR 73.310]

filing date

The date upon which a document must be filed after all computations of time authorized by this section have been made. [47 CFR 1.4]

filing period

The number of days allowed or prescribed by statute, rule, order, notice or other Commission action for filing any document with the Commission. It does not include any additional days allowed for filing any document pursuant to paragraphs (g), (h), and (j) of this section. [47 CFR 1.4]

fill-in area

The area where the coverage contour of an FM translator or booster station is within the protected contour of the associated primary station (*i.e.*, predicted 0.5 mV/m contour for commercial Class B stations, predicted 0.7 mV/m contour for commercial Class B1 stations, and predicted 1 mV/m contour for all other classes of stations). [47 CFR 74.1201]

fill-in transmitters

Transmitters added to a station, in the same area and transmitting on the same channel or channel block as previously authorized transmitters, that do not expand the existing service area, but are established for the purpose of improving reception in dead spots. [47 CFR 22.99]

filtering

Refers to the requirement in §95.633(b). [47 CFR 95.669]

FIPS

Federal Information Processing Standard [47 CFR 11.31]

five-year build-out period

A five-year period during which the licensee of the first cellular system authorized on each channel block in each cellular market may expand the system within that market. See §22.947. [47 CFR 22.99]

fixed earth station

An earth station intended to be used at a specified fixed point. [47 CFR 21.2, 25.201]

fixed microwave auxiliary station

A fixed station used in connection with (1) the alignment of microwave transmitting and receiving antenna systems and equipment, (2) coordination of microwave radio survey operations, and (3) cue and contact control of television pickup station operations. [47 CFR 21.2]

fixed public press service

A limited radio communication service carried on between point-to-point telegraph stations, consisting of transmissions by fixed stations open to limited public correspondence, of news items, or other material related to or intended for publication by press agencies, newspapers, or for public dissemination. In addition, these transmissions may be directed to one or more fixed points specifically named in a station license, or to unnamed points in accordance with the provisions of §23.53. [47 CFR 23.1]

fixed public service

A radiocommunication service carried on between fixed stations open to public correspondence. [47 CFR 23.1]

fixed relay station [1]

A station at a specified site used to communicate with another station at another specified site. [47 CFR 90.7]

fixed relay station [2]

An operational-fixed station associated with one or more stations in the mobile service, established to receive radio signals directed to it and to retransmit them automatically on a fixed service frequency. [47 CFR 94.3]

fixed satellite service

A radiocommunication service between earth stations at given positions, when one or more satellites are used; the given position may be a specified fixed point or any fixed point within specified areas; in some cases this service includes satellite-to-satellite links, which may also be operated in the inter-satellite service; the fixed-satellite service may also include feeder links for other space radiocommunication services. (RR) [47 CFR 2.1, 25.201]

fixed service

A radiocommunication service between specified fixed points. (RR) [47 CFR 2.1, 5.4, 24.5, 26.4, 94.3]

fixed station [1]

A station in the fixed service. (RR) [47 CFR 2.1, 5.4]

fixed station [2]

A station at a fixed location. [47 CFR 21.2]

fixed station [3]

In the fixed public or fixed public press service includes all apparatus used in rendering the authorized service at a particular location under a single instrument of authorization. [47 CFR 23.1]

fixed transmitter

A stationary transmitter that communicates with other stationary transmitters. [47 CFR 22.99]

FLA

Flood watch [47 CFR 11.31]

fleet radio station license

An authorization issued by the Commission for two or more ships having a common owner or operator. [47 CFR 80.5]

flight test aircraft station

An aircraft station used in the testing of aircraft or their major components. [47 CFR 87.5]

flight test land station

An aeronautical station used in the testing of aircraft or their major components. [47 CFR 87.5]

FLS

Flood statement [47 CFR 11.31]

FLW

Flood warning [47 CFR 11.31]

FM

Frequency modulation [47 CFR 73.681]

FM blanketing

Blanketing is that form of interference to the reception of other broadcast stations which is caused by the presence of an FM broadcast signal of 115 dBu (562 mV/m) or greater signal strength in the area adjacent to the antenna of the transmitting station. The 115 dBu contour is referred to as the blanketing contour and the area within this contour is referred to as the blanketing area. [47 CFR 73.310]

FM broadcast band

The band of frequencies extending from 88 to 108 MHZ, which includes those assigned to non-commercial educational broadcasting. [47 CFR 73.310]

FM broadcast booster station

A station in the broadcasting service operated for the sole purpose of retransmitting the signals of an FM radio broadcast station, by amplifying and reradiating such signals, without significantly altering any characteristic of the incoming signal other than its amplitude. [47 CFR 74.1201]

FM broadcast channel

A band of frequencies 200 kHz wide and designated by its center frequency. Channels for FM broadcast stations begin at 88.1 MHZ and continue in successive steps of 200 kHz to and including 107.9. [47 CFR 73.310]

FM broadcast station

A station employing frequency modulation in the FM broadcast band and licensed primarily for the transmission of radio-telephone emissions intended to be received by the general public. [47 CFR 73.310]

FM radio broadcast station

When used in the Subpart L, the term FM broadcast station or FM radio broadcast station refers to commercial and noncommercial educational FM radio broadcast stations as defined in §2.1 of this chapter, unless the context indicates otherwise. [47 CFR 74.1201]

FM stereophonic broadcast

The transmission of a stereophonic program by a single FM broadcast station utilizing the main channel and a stereophonic subchannel. [47 CFR 73.310]

FM translator

A station in the broadcasting service operated for the purpose of retransmitting the signals of an FM radio broadcast station or another FM broadcast translator station without significantly altering any characteristics of the incoming signal other than its frequency and amplitude, in order to provide FM broadcast service to the general public. [47 CFR 74.1201]

FOB

Field Operations Bureau [47 CFR 0.445]

FOIA

Freedom of Information Act [47 CFR 0.461]

foreign communication

Communication or transmission from or to any place in the United States to or from a foreign country, or between a station in the United States and a mobile station located outside the United States. [47 USC 153]

foreign government

Any sovereign empire, kingdom, state, or independent political community, including foreign diplomatic and consular establishments and coalitions or associations of governments (*e.g.*, North Atlantic Treaty Organization (NATO), Southeast Asian Treaty Organization (SEATO), Organization of American States (OAS), and government agencies or organization (*e.g.*, Pan American Union, International Postal Union, and International Monetary Fund)). [47 CFR 64.1604]

formal opposition or formal complaint

(1) A pleading opposing the grant of a particular application, waiver request, petition for special relief or other request for Commission action, or a pleading in the nature of a complaint (other than a section 208 complaint), which meets the following requirements (i) the caption and text of a pleading make it unmistakably clear that the pleading is intended to be a formal opposition or formal complaint; (ii) the pleading is served upon the other parties to the proceeding or, in the case of a complaint, upon the person subject to the complaint; and (iii) the pleading is filed within the time period, if any, prescribed for such a pleading;

(2) A formal complaint under section 208 of the Communications Act if it meets the requirements of §1.721 *et seq.* of the Commission's Rules (formal complaints against common carriers). [47 CFR 1.1202]

forward links

Transmissions in the frequency bands specified in §90.357(a) and used to control and interrogate the mobile units to be located by multialteration LMS systems. [47 CFR 90.7]

FPLMTS

Future public land mobile telecommunication systems [47 CFR 2.106]

FR

Federal regulation [47 CFR 1.1108]

frame

Scanning all of the picture area once. In the line interlaced scanning pattern of two to one, a frame consists of two fields. [47 CFR 73.681]

franchise

An initial authorization, or renewal thereof (including a renewal of an authorization which has been granted subject to Section 626), issued by a franchising authority, whether such authorization is designated as a franchise, permit, license, resolution, contract, certificate, agreement, or otherwise, which authorizes the construction or operation of a cable system. [47 USC 522]

franchise expiration

The date of the expiration of the term of the franchise, as provided under the franchise, as it was in effect on the date of the enactment of this title. [47 USC 546]

franchisee fee

Includes any tax, fee, or assessment of any kind imposed by a franchising authority or other government entity on a cable operator or cable subscriber, or both, solely because of their status as such.

The term does not include--

(A) any tax, fee, or assessment of general applicability (including any such tax, fee, or assessment imposed on both utilities and cable operators or their services but not including a tax, fee, or assessment which is unduly discriminatory against cable operators or cable subscribers);

(B) in the case of any franchise in effect on the date of the enactment of this title, payments which are required by the franchise to be made by the cable operator during the term of such franchise for, or in support of the use of, public, educational, or governmental access facilities;

(C) in the case of any franchise granted after such date of enactment, capital costs which are required by the franchise to be incurred by the cable operator for public, educational, or governmental access facilities;

(D) requirements or charges incidental to the awarding or enforcing of the franchise, including payments for bonds, security funds, letters of credit, insurance, indemnification, penalties, or liquidated damages; or

(E) any fee imposed under Title 17, United States Code. [47 USC 542]

franchising authority

Any governmental entity empowered by federal, state, or local law to grant a franchise. [47 USC 522]

frank
Any authority which authorizes free, or partially free, service. [47 CFR 41.1]

free space field strength
The field strength that would exist at a point in the absence of waves reflected from the earth or other reflecting objects. [47 CFR 73.310, 73.681]

frequency
The number of cycles occurring per second of an electrical or electromagnetic wave; a number of [sic] representing a specific point in the electromagnetic spectrum. [47 CFR 22.99]

frequency coordinator [1]
An entity or organization that has been certified by the Commission to recommend frequencies for use by licensees in the Private Land Mobile Radio Services. [47 CFR 90.7]

frequency coordinator [2]
An entity, recognized in a local or regional area by amateur operators whose stations are eligible to be auxiliary or repeater stations, that recommends transmit/receive channels and associated operating and technical parameters for such stations in order to avoid or minimize potential interference. [47 CFR 97.3]

frequency departure
The amount of variation of a carrier frequency or center frequency from its assigned value. [47 CFR 73.14, 73.310, 73.681]

frequency deviation [1]
The peak difference between modulated wave and the carrier frequency. [47 CFR 73.310]

frequency deviation [2]
The peak difference between the instantaneous frequency of the modulated wave and the carrier frequency. [47 CFR 73.681]

frequency hopping systems

A frequency hopping system is a spread spectrum system in which the carrier is modulated with the coded information in a conventional manner causing a conventional spreading of the RF energy about the carrier frequency. However, the frequency of the carrier is not fixed but changes at fixed intervals under the direction of a pseudorandom code sequence. The wide RF bandwidth needed by such a system is not required by a spreading of the RF energy about the carrier but rather to accommodate the range of frequencies to which the carrier frequency can hop. [47 CFR 2.1]

frequency-hour

One frequency used for one hour regardless of the number of transmitters over which it is simultaneously broadcast by a station during that hour. [47 CFR 73.701]

frequency modulation (FM)

A system of modulation where the instantaneous radio frequency varies in proportion to the instantaneous amplitude of the modulating signal (amplitude of modulating signal to be measured after pre-emphasis, if used) and the instantaneous radio frequency is independent of the frequency of the modulating signal. [47 CFR 73.310, 73.681]

frequency-shift telegraphy

Telegraphy by frequency modulation in which the telegraph signal shifts the frequency of the carrier between predetermined values. (RR) [47 CFR 2.1]

frequency swing

The peak difference between the maximum and the minimum values of the instantaneous frequency of the carrier wave during modulation. [47 CFR 73.310, 73.681]

frequency tolerance [1]

The maximum permissible departure by the centre frequency of the frequency band occupied by an emission from the assigned frequency or, by the characteristic frequency of an emission from the reference frequency. [47 CFR 2.1, 23.1] Note: The frequency tolerance is expressed in parts in 10 to the sixth power or in Hertz. (RR) [47 CFR 21.2]

frequency tolerance [2]

The maximum permissible variation of the carrier frequency expressed as a percentage or in Hertz.

frequency tolerance [3]

The maximum permissible departure with respect to the assigned frequency of the corresponding characteristic frequency of an emission. For purposes of this part, the frequency tolerance is expressed as a percentage of the assigned frequency. [47 CFR 94.3]

full carrier single-sideband emission

A single-sideband emission without suppression of the carrier. (RR) [47 CFR 2.1]

full network station

A commercial television broadcast station that generally carries in weekly prime time hours 85 percent of the hours of programing offered by one of the three major national television networks with which it has a primary affiliation *(i.e.,* right of first refusal or first call). [47 CFR 76.5]

fully protected nonsystem premises wiring

Non-system premises wiring which is electrically behind registered (or grandfathered) equipment or protective circuitry which assures that electrical contact between the wiring and commercial power wiring or earth ground will not result in hazardous voltages at the telephone network interface. [47 CFR 68.3]

fully protected systems premises wiring

Premises wiring which is either:

(1) No greater than 15 meters (50 feet) in length (measured linearly between the points where it leaves equipment or connector housings) and registered as a component of and supplied to the user with the registered terminal equipment or protective circuitry with which it is to be used. Such wiring shall either be pre-connected to the equipment or circuitry, or may be so connected by the user (or others) it is demonstrated in the registration application that such connection by the untrained will not result in harm, using relatively failsafe means.

(2) A cord which complies with the previous subsection either as an integral length or in combination with no more than one connectorized

extension cord. If used, the extension cord must comply with the requirements of §68.200(h) of these Rules.

(3) Wiring located in an equipment room with restricted access, provided that this wiring remains exposed for inspection and is not concealed or embedded in the building's structure, and that it conforms to §68.215(d).

(4) Electrically behind registered (or grandfathered) equipment, system components or protective circuitry components or protective circuitry which assure that electrical contact between the wiring and commercial power wiring or earth ground will not result in hazardous voltages or excessive longitudinal imbalance at the telephone network interface. [47 CFR 68.3]

Fund

The Telecommunications Development Fund established pursuant to this section. [47 USC 614]

G

G

Government [47 CFR 2.106]

GAAP

Generally accepted accounting principles [47 CFR 24.720]

gain of an antenna

The ratio, usually expressed in decibels, of the power required at the input of a loss free reference antenna to the power supplied to the input of the given antenna to produce, in a given direction, the same field strength or the same power fluxdensity at the same distance. When not specified otherwise, the gain refers to the direction of maximum radiation. The gain may be considered for a specified polarization. Note: Depending on the choice of the reference antenna a distinction is made between (1) absolute or isotropic gain (Gi), when the reference antenna is an isotropic antenna isolated in space; (2) gain relative to a half-wave dipole (Gd), when the reference antenna is a half-wave dipole isolated in space whose equatorial plane contains the given direction; (3) gain relative to a short vertical antenna (Gv), when the reference antenna is a linear conductor, much shorter than one quarter of the wavelength, normal to the surface of a perfectly conducting plane which contains the given direction. (RR) [47 CFR 2.1]

GDP Price Index (GDP-PI)

The estimate of the "Fixed Weight Price Index for Gross Domestic Product, 1987 Weights" published by the United States Department of Commerce, which the Commission designates by Order. [47 CFR 61.3]

general communication

Two-way voice communication, through a base station, between (1) a common carrier land mobile or airborne station and a landline telephone station connected to a public message landline telephone system, or (2) two common carrier land mobile stations, or (3) two common carrier airborne stations, or (4) a common carrier land mobile station and a common carrier airborne station. [47 CFR 21.2]

general purpose mobile service

A mobile service that includes all mobile communications uses including those within the Aeronautical Mobile, Land Mobile, or the Maritime Mobile Services. [47 CFR 2.1]

general support facilities

Include buildings, land, vehicles, aircraft, work equipment, furniture, office equipment and general purpose computers as described in the *Separations Manual* and included in Account 2110. [47 CFR 69.2]

generic requirement

A description of acceptable product attributes for use by local exchange carriers in establishing product specifications for the purchase of telecommunications equipment, customer premises equipment, and software integral thereto. [47 USC 273]

geographic center

The geographic center of an urbanized area is defined by the coordinates given at Table 1 of §90.635. [47 CFR 90.7]

geostationary satellite

A geosynchronous satellite whose circular and direct orbit lies in the plane of the earth's equator and which thus remains fixed relative to the earth; by extension, a satellite which remains approximately fixed relative to the earth. (RR) [47 CFR 2.1, 25.201]

geostationary satellite orbit

The orbit in which a satellite must be placed to be a geostationary satellite. (RR) [47 CFR 2.1]

geosynchronous satellite

An earth satellite whose period of revolution is equal to the period of rotation of the earth about its axis. (RR) [47 CFR 2.1]

GHz

Gigahertz [47 CFR 2.106]

gift

Any unconditional gift, donation or bequest of real, personal and other property (including voluntary and uncompensated services as authorized under 5 U.S.C. 3109). [47 CFR 1.3001]

glide path station

A radionavigation land station which provides vertical guidance to aircraft during approach to landing. [47 CFR 87.5]

global maritime distress and safety system (GMDSS)

An International Maritime Organization (IMO) worldwide coordinated maritime distress system designed to provide the rapid transfer of distress messages from vessels in distress to units best suited for giving or coordinating assistance. The system includes standardized equipment and operational procedures, unique identifiers for each station, and the integrated use of frequency bands and radio systems to ensure the transmission and reception of distress and safety calls and messages at short, medium and long ranges. [47 CFR 80.5]

GMDSS

Global Maritime Distress and Safety System [47 CFR 2.106]

GMRS

General mobile radio service [47 CFR 95.669]

GMRS system

(a) One or more transmitting units used by station operators to communicate messages. A GMRS system is comprised of (1) one or more station operators; (2) one mobile station consisting of one or more mobile units (see §95.23); (3) one or more land stations (optional); and (4) paging receivers (optional).

(b) In certain areas, point-to-point GMRS systems may be comprised of fixed stations only (see §§95.47, 95.49 and 95.61).

(c) A GMRS system may be operated in (1) simplex mode—only one station operator can speak at a time; (2) duplex mode—two station operators can speak at the same time. One or more stations transmit on one channel.

The other station(s) transmit(s) on the channel pair counterpart; (3) combined simplex-duplex mode—a mobile relay system with mobile units operating in simplex mode on a channel pair. [47 CFR 95.21]

GMRS transmitter

A transmitter that operates or is intended to operate at a station authorized in the GMRS. [47 CFR 95.669]

GNP Price Index (GNP-PI)

The estimate of the "Fixed-Weighted Price Index for Gross National Product, 1982 Weights" published by the United States Department of Commerce, which the Commission designates by Order. [47 CFR 61.3]

government

The federal government or any foreign, state, county, municipal, or other local government agency or organization. Specific qualifications will be supplied whenever reference to a particular level of government is intended (*e.g.*, "federal government," "state government"). "Foreign government" means any non-U.S. sovereign empire, kingdom, state, or independent political community, including foreign diplomatic and consular establishments and coalitions or associations of governments (*e.g.*, North Atlantic Treaty Organization (NATO), Organization of American States (OAS), and United Nations (UN); and associations of governments or government agencies or organizations (*e.g..*, Pan American Union, International Postal Union, and International Monetary Fund). [47 CFR Pt. 216, App., 47 CFR 64.1604]

GPO

U.S. Government Printing Office [47 CFR 0.409]

grade A and grade B contours

The field intensity contours defined in §73.683(a) of this chapter. [47 CFR 76.5]

grade B contour

The field strength of a television broadcast station computed in accordance with regulations promulgated by the Commission. [47 USC 522]

Great Lakes

As used in this part in reference to the Great Lakes Radio Agreement, means all of Lakes Ontario, Erie, Huron (including Georgian Bay), Michigan, Superior, their connecting and tributary waters and the St. Lawrence River as far east as the lower exit of the St. Lambert Lock as Montreal in the Province of Quebec, Canada, but does not include any connecting and tributary waters other than: the St. Mary's River, the St. Clair River, Lake St. Clair, the Detroit River and the Welland Canal. [47 CFR 80.5]

Great Lakes Agreement

The Agreement for the Promotion of Safety on the Great Lakes by Means of Radio in force and the regulations referred to therein. [47 USC 153]

gross revenues [1]

All income received by an entity, whether earned or passive, before any deductions are made for costs of doing business (*e.g.*, cost of goods sold), as evidenced by audited financial statements for the preceding relevant number of calendar years, or, if audited financial statements were not prepared on a calendar-year basis, for the preceding relevant number of fiscal years. If an entity was not in existence for all or part of the relevant period, gross revenues shall be evidenced by the audited financial statements of the entity's predecessor-in-interest or, if there is no identifiable predecessor-in-interest, unaudited financial statements certified by the applicant as accurate. [47 CFR 21.961]

gross revenues [2]

All income received by an entity, whether earned or passive, before any deductions are made for costs of doing business (*e.g.*, cost of goods sold), as evidenced by audited quarterly financial statements for the relevant period. [47 CFR 24.320]

gross revenues [3]

All income received by an entity, whether earned or passive, before any deductions are made for costs of doing business (*e.g.*, cost of goods sold), as evidenced by audited financial statements for the relevant number of calendar years preceding January 1, 1994, or, if audited, financial statements were not prepared on a calendar-year basis, for the most recently completed fiscal years preceding the filing of the applicant's short-form application (Form 175). For short-form applications filed after December 31, 1995, gross revenues shall be evidenced by audited financial statements for the preceding relevant number of calendar or fiscal years. If an entity was not in existence for all or part of the relevant period, gross revenues shall be evidenced by the audited financial statements of the entity's predecessor-in-interest or, if there is no identifiable predecessor-in-interest, unaudited financial statements certified by the applicant as accurate. [47 CFR 24.720, 26.4]

gross revenues [4]

For applications filed after December 31, 1994, gross revenues shall be evidenced by audited financial statements for the preceding relevant number of calendar or fiscal years. If an entity was not in existence for all or part of the relevant period, gross revenues shall be evidenced by the audited financial statements of the entity's predecessor-in-interest or, if there is no identifiable predecessor-in-interest, unaudited financial statements certified by the applicant as accurate. [47 CFR 90.814]

ground station

In the Air-Ground Radiotelephone Service, a stationary transmitter that provides service to airborne mobile stations. [47 CFR 22.99]

group plan

As applied to depreciation accounting, means the plan under which depreciation charges are accrued upon the basis of the original cost of all property included in each depreciable plant account, using the average service life thereof properly weighted, and upon the retirement of any depreciable property its cost is charged to the depreciation reserve whether or not the particular item has attained the average service life. [47 CFR 32.9000]

GTE Consent Decree

The order entered December 21, 1984, as restated January 11, 1985, in the action styled United States v. GTE Corp., Civil Action No. 83-1298, in the United States District Court for the District of Columbia, and any judgment or order with respect to such action entered on or after December 21, 1984. [47 USC 601]

GWCS

General wireless communications service [47 CFR 26.1]

H

harbor (port)
Any place to which ships may resort for shelter or to load or unload passengers or goods, or to obtain fuel, water, or supplies. This term shall apply to such places whether proclaimed public or not and whether natural or artificial. [47 USC 153, 47 CFR 80.5]

harm
Electrical hazards to telephone company personnel, damage to telephone company equipment, malfunction of telephone company billing equipment, and degradation of service to persons other than the user of the subject terminal equipment, his calling or called party. [47 CFR 68.3]

harmful interference [1]
Interference which endangers the functioning of a radionavigation service or of other safety services or seriously degrades, obstructs, or repeatedly interrupts a radiocommunication service operating in accordance with these [international] Radio Regulations. (RR) [47 CFR 2.1, 97.3, 18.107]

harmful interference [2]
Any radiation or any induction which endangers the functioning of a radio-navigation service or of a safety service or obstructs or repeatedly interrupts a radio service [47 CFR 21.2] operating in accordance with the Table of Frequency Allocations and other provisions of part 2 of this chapter. [47 CFR 5.4]

harmful interference [3]
Any emission, radiation or induction that endangers the functioning of a radio navigation service or of other safety services or seriously degrades, obstructs or repeatedly interrupts a radiocommunications service operating in accordance with this chapter. [47 CFR 15.3]

harmful interference [4]

For the purposes of resolving conflicts between stations operating under this part, any emission, radiation, or induction which specifically degrades, obstructs, or interrupts the service provided by such stations. [47 CFR 90.7]

harmful interference [5]

Any transmission, radiation or induction that endangers the functioning of a radionavigation or other safety service or seriously degrades, obstructs or repeatedly interrupts a radiocommunication service operating in accordance with applicable laws, treaties and regulations. [47 CFR 95.669]

HBO

Home Box Office [47 CFR 11.43]

HCO

Hearing carry over [47 CFR 64.601]

health care provider

(i) Post-secondary educational institutions offering health care instruction, teaching hospitals, and medical schools; (ii) community health centers or health centers providing health care to migrants; (iii) local health departments or agencies; (iv) community mental health centers; (v) not-for-profit hospitals; (vi) rural health clinics; and (vii) consortia of health care providers consisting of one or more entities described in clauses (i) through (vi). [47 USC 254]

hearing aid

A personal electronic amplifying device, intended to increase the loudness of sound and worn to compensate for impaired hearing. When equipped with an optional inductive pick-up coil (commonly called a telecoil), a hearing aid can be used to amplify magnetic fields such as those from telephone receivers or induction-loop systems. [47 CFR 68.316]

hearing carry over (HCO)

A reduced form of TRS where the person with the speech disability is able to listen to the other end user and, in reply, the CA speaks the text as typed by the person with the speech disability. The CA does not type any conversation. [47 CFR 64.601]

height above average terrain (HAAT)

The height of an antenna above the average elevation of the surrounding area. [47 CFR 22.99]

HF

high frequency [47 CFR 97.3]

high definition television

Systems that offer approximately twice the vertical and horizontal resolution of receivers generally available on the date of enactment of the Telecommunications Act of 1996, as further defined in the proceedings described in paragraph (1) of this subsection. [47 USC 335]

high frequency (HF)

The frequency range 3-30 kHz. [47 CFR 97.3]

HLS

Hurricane statement [47 CFR 11.31]

holding time

The time in which an item of telephone plant is in actual use either by a customer or an operator. For example, on a completed telephone call, holding time includes conversation time as well as other time in use. At local dial offices any measured minutes which result from other than customer attempts to place calls (as evidenced by the dialing of at least one digit) are not treated as holding time. [47 CFR Pt. 36, App.]

holiday

Saturday, Sunday, officially recognized federal legal holidays and any other day on which the Commission's offices are closed and not reopened prior to 5:30 p.m. For example, a regularly scheduled Commission business day may become a holiday if its offices are closed prior to 5:30 p.m. due to adverse weather, emergency or other closing. [47 CFR 1.4]

host central office

An electronic analog or digital base switching unit containing the central call processing functions which service the host office and its remote locations. [47 CFR Pt. 36, App.]

HUA

Hurricane watch [47 CFR 11.31]

HUW

Hurricane warning [47 CFR 11.31]

HWA

High wind watch [47 CFR 11.31]

HWW

High wind warning [47 CFR 11.31]

hybrid spread spectrum systems

Hybrid spread spectrum systems are those which use combinations of two or more types of direct sequence, frequency hopping, time hopping and pulsed FM modulation in order to achieve their wide occupied bandwidths. [47 CFR 2.1]

Hz

Hertz [47 CFR 97.3]

I

ICAO
> Convention on International Civil Aviation [47 CFR 87.1]

ICR
> Inter-city relay [47 CFR 74.501]

IEEE
> Institute of Electrical and Electronics Engineers, Inc. [47 CFR 24.237]

IF
> Intermediate frequency [47 CFR 73.203]

immediate family member
> Father, mother, husband, wife, son, daughter, brother, sister, father- or mother-in-law, son- or daughter-in-law, brother- or sister-in-law, step-father or -mother, step-brother or -sister, step-son or -daughter, half brother or sister. This presumption may be rebutted by showing that the family members are estranged, the family ties are remote, or the family members are not closely involved with each other in business matters. [47 CFR 1.2110]

IMO
> International Maritime Organization [47 CFR 80.5]

impairment of service (see discontinuance of service)

inband signaling private line interface

The point of connection between an inband signaling voiceband private line and terminal equipment or systems where the signaling frequencies are within the voiceband. All tip and ring leads shall be treated as telephone connections for the purposes of fulfilling registration conditions. [47 CFR 68.3]

inbound telemarketing

The marketing of property, goods, or services by telephone to a customer or potential customer who initiated the call. [47 USC 274]

in-building radiation systems

Supplementary systems comprising low power transmitters, receivers, indoor antennas and/or leaky coaxial cable radiators, designed to improve service reliability inside buildings or structures located within the service areas of stations in the Public Mobile Services. [47 CFR 22.99]

incidental interLATA services

The interLATA provision by a Bell operating company or its affiliate—

(1)(A) of audio programming, video programming, or other programming services to subscribers to such services of such company or affiliate; (B) of the capability for interaction by such subscribers to select or respond to such audio programming, video programming, or other programming services; (C) to distributors of audio programming or video programming that such company or affiliate owns or controls, or is licensed by the copyright owner of such programming (or by an assignee of such owner) to distribute; or (D) of alarm monitoring services;

(2) of two-way interactive video services or Internet services over dedicated facilities to or for elementary and secondary schools as defined in Section 254(h)(5);

(3) of commercial mobile services in accordance with Section 332(C) of this Act and with the regulations prescribed by the Commission pursuant to paragraph (8) of such section;

(4) of a service that permits a customer that is located in one LATA to retrieve stored information from, or file information for storage in, information storage facilities of such company that are located in another LATA;

(5) of signaling information used in connection with the provision of telephone exchange services or exchange access by a local exchange carrier; or

(6) of network control signaling information to, and receipt of such signaling information from, common carriers offering interLATA services at any location within the area in which such Bell operating company provides telephone exchange services or exchange access. [47 USC 271]

incidental phase modulation

The peak phase deviation (in radians) resulting from the process of amplitude modulation. [47 CFR 73.14]

incidental radiator

A device that generates radio frequency energy during the course of its operation although the device is not intentionally designed to generate or emit radio frequency energy. Examples of incidental radiators are dc motors, mechanical light switches, etc. [47 CFR 15.3]

inclination of an orbit (of an earth satellite)

The angle determined by the plane containing the orbit and the plane of the earth's equator. (RR) [47 CFR 2.1]

incumbent

An MDS station that was authorized or proposed before September 15, 1995, including those stations that are subsequently modified, renewed or reinstated. [47 CFR 21.2]

incumbent local exchange carrier

With respect to an area, the local exchange carrier that (A) on the date of enactment of the Telecommunications Act of 1996, provided telephone exchange service in such area; and (B)(i) on such date of enactment, was deemed to be a member of the exchange carrier association pursuant to section 69.601(b) of the Commission's regulations (47 CFR 69.601(b)); or (ii) is a person or entity that, on or after such date of enactment, became a successor or assign of a member described in clause (i). [47 USC 251]

independent station

A commercial television broadcast station that generally carries in prime time not more than 10 hours of programming per week offered by the three major national television networks. [47 CFR 76.5]

independent syndicator

One not owned or controlled, in full or in part, by a television network. [47 CFR 73.662]

indicator

Words, letters or numerals appended to and separated from the call sign during the station identification. [47 CFR 97.3]

individual [1]

A citizen of the United States or an alien lawfully admitted for permanent residence. [47 CFR 0.551]

individual [2]

A natural person [47 CFR 1.1918]

individual with handicaps

Any individual who has a physical or mental impairment that substantially limits one or more major life activities, has a record of such an impairment, or is regarded as having such an impairment. As used in this definition, the phrase:

(a) *physical or mental impairment* includes (i) any physiological disorder or condition, cosmetic disfigurement, or anatomical loss affecting one or more of the following body systems: Neurological; musculoskeletal; special sense organs; respiratory, including speech organs; cardiovascular; reproductive; digestive; genitourinary; hemic and lymphatic; skin; and endocrine; or (ii) any mental or psychological disorder, such as mental retardation, organic brain syndrome, emotional or mental illness, and specific learning disabilities. The term "physical or mental impairment" includes, but is not limited to, such diseases and conditions as orthopedic, visual, speech, and hearing impairments, cerebral palsy, epilepsy, muscular dystrophy, multiple sclerosis, cancer, heart disease, diabetes, mental retardation, emotional illness, and drug addiction and alcoholism.

(b) *Major life activities* includes functions such as caring for one's self, performing manual tasks, walking, seeing, hearing, speaking, breathing, learning, and working.

(c) *Has a record of such an impairment* means has a history of, or has been misclassified as having, a mental or physical impairment that substantially limits one or more major life activities.

(d) *Is regarded as having an impairment* (i) has a physical or mental impairment that does not substantially limit major life activities but is treated by the Commission as constituting such a limitation; (ii) has a physical or mental impairment that substantially limits major life activities only as a result of the attitudes of others toward such impairment; or (iii) has none of the impairments defined in paragraph (1) of this definition but is treated by the Commission as having such an impairment. [47 CFR 1.1803]

individual reception (in the broadcasting-satellite service)

The reception of emissions from a space station in the broadcasting-satellite service by simple domestic installations and in particular those possessing small antennae. (RR) [47 CFR 2.1]

industrial heating equipment

A category of ISM equipment used for or in connection with industrial heating operations utilized in a manufacturing or production process. [47 CFR 18.107]

industrial, scientific and medical (ISM) (of radio frequency energy) applications

Operations of equipment or appliances designed to generate and use locally radio-frequency energy for industrial, scientific, medical, domestic or similar purposes, excluding applications in the field of telecommunications. (RR) [47 CFR 2.1]

industrial, scientific, and medical (ISM) equipment

Equipment or appliances designed to generate and use locally RF energy for industrial, scientific, medical, domestic or similar purposes, excluding applications in the field of telecommunication. Typical ISM applications are the production of physical, biological, or chemical effects such as heating, ionization of gases, mechanical vibrations, hair removal and acceleration charged particles. [47 CFR 18.107]

industry-wide

Activities funded by or performed on behalf of local exchange carriers for use in providing wireline telephone exchange service whose combined total of deployed access lines in the United States constitutes at least 30 percent of all access lines deployed by telecommunications carriers in the United States as of the date of enactment of the Telecommunications Act of 1996. [47 USC 273]

information bulletin

A message directed only to amateur operators consisting solely of subject matter of direct interest to the amateur service. [47 CFR 97.3]

information content provider

Any person or entity that is responsible, in whole or in part, for the creation or development of information provided through the Internet or any other interactive computer service. [47 USC 230]

information origination/termination equipment [1]

Equipment used to input into or receive output from the telecommunications network. [47 CFR Pt. 36, App.]

information origination/termination equipment [2]

Includes all equipment or facilities that are described as information origination/termination equipment in the *Separations Manual* and in Account 2310 except information origination/termination equipment that is used by telephone companies in their own operations. [47 CFR 69.2]

information service

The offering of a capability for generating, acquiring, storing, transforming, processing, retrieving, utilizing, or making available information via telecommunications, and includes electronic publishing, but does not include any use of any such capability for the management, control, or operation of a telecommunications system or the management of a telecommunications service. [47 USC 153]

initial cellular applications

Applications for authority to construct and operate a new cellular system, excluding applications for interim operating authority. [47 CFR 22.99]

initial transport rates

Rates for entrance facilities, direct-trunked transport, tandem-switched transport, dedicated signalling transport, and the interconnection charge in tariffs filed on September 1, 1993 pursuant to the Report and Order in Transport Rate Structure and Pricing, CC Docket No. 91-213, FCC 92-442, 7 FCC Rod 7006 (1992). [47 CFR 69.2]

inland waters [1]

As used in reference to waters of the United States, its territories and possessions, waters that lie landward of the boundary lines of inland waters as contained in 33 CFR part 82, as well as waters within its land territory, such as rivers and lakes, over which the United States exercises sovereignty. [47 CFR 80.5]

inland waters [2]

Inland waters of western Washington and British Columbia bounded by 47 degrees latitude on the south, the Canada/U.S.A. Coordination Zone Line B on the north, and to the west by 124 degrees 40 minutes longitude at the west entrance to the Strait of Juan de Fuca. [47 CFR 80.57]

inland waters local channel

A channel designed to provide local coverage of certain bays, inlets and ports where coverage by primary or supplementary channels is poor or where heavy traffic loading warrants. A local channel must not cause harmful interference to any primary or supplementary channels. Coverage must be confined to the designated sector. [47 CFR 80.57]

inland waters primary channel

A channel intended to cover the greater portion of an Inland Waters Public Correspondence Sector. It may provide some coverage to an adjacent sector but must not provide coverage beyond the adjacent sector. Harmful interference beyond the adjacent sector must not occur. Only one primary channel will be authorized in any sector. [47 CFR 80.57]

inland waters public correspondence sector

A distinct geographical area in which one primary and one supplementary channel is allotted. A number of local channels may also be authorized. [47 CFR 80.57]

inland waters supplementary channel
    A channel intended to improve coverage within a sector or to relieve traffic congestion on the primary channel. It may provide some coverage of an adjacent sector but must not provide coverage beyond the adjacent sector Harmful interference beyond the adjacent sector must not occur. Only one supplementary channel will be authorized in any sector. [47 CFR 80.57]

INMARSAT
    International Maritime Satellite service [47 CFR 0.453]

input power
    The product of the direct voltage applied to the last radio stage and the total direct current flowing to the last radio stage, measure without modulation. [47 CFR 73.14]

input selector switch
    Any device that enables a viewer to select between cable service and off-the-air television signals. Such a device may be more sophisticated than a mere two-sided switch, may utilize other cable interface equipment, and may be built into consumer television receivers. [47 CFR 76.5]

in-region state
    A state in which a Bell operating company or any of its affiliates was authorized to provide wireline telephone exchange service pursuant to the reorganization plan approved under the AT&T Consent Decree, as in effect on the day before the date of enactment of the Telecommunications Act of 1996. [47 USC 271]

institution of higher education
    Has the meaning provided in Section 1201 of the Higher Education Act of 1965 (20 USC 1141). [47 USC 223]

institutional investor
    An insurance company, a bank holding stock in trust accounts through its trust department, or an investment company as defined in 15 U.S.C. 80a-3(a), including within such definition any entity that would otherwise meet the definition of investment company under 15 U.S.C. 80a-3(a) but is excluded by the exemptions set forth in 15 U.S.C. 80a-3 (b) and (c), without regard to whether such entity is an issuer of securities; provided that, if such investment

company is owned, in whole or in part, by other entities, such investment company, such other entities and the affiliates of such other entities, taken as a whole, must be primarily engaged in the business of investing, reinvesting or trading in securities or in distributing or providing investment management services for securities. [47 CFR 24.720]

instructional television fixed station
A fixed station operated by an educational organization and used primarily for the transmission of visual and aural instructional, cultural, and other types of educational material to one or more fixed receiving locations. [47 CFR 74.901]

instrument landing system (ILS)
A radionavigation system which provides aircraft with horizontal and vertical guidance just before and during landing and, at certain fixed points, indicates the distance to the reference point of landing. (RR) [47 CFR 2.1, 87.5]

instrument landing system glide path
A system of vertical guidance embodied in the instrument landing system which indicates the vertical deviation of the aircraft from its optimum path of descent. (RR) [47 CFR 2.1, 87.5]

instrument landing system localizer
A system of horizontal guidance embodied in the instrument landing system which indicates the horizontal deviation of the aircraft form its optimum path of descent along the axis of the runway. (RR) [47 CFR 2.1, 87.5]

intangible property
Assets that have no physical existence but instead have value because of the rights which ownership confers. [47 CFR 32.9000]

INTELSAT
International Telecommunications Satellite Organization [47 CFR 0.453]

intentional radiator
A device that intentionally generates and emits radio frequency energy by radiation or induction. [47 CFR 15.3]

interactive computer service
Any information service, system, or access software provider that provides or enables computer access by multiple users to a computer server, including specifically a service or system that provides access to the Internet and such systems operated or services offered by libraries or educational institutions. [47 USC 230]

interactive on-demand services
A service providing video programming to subscribers over switched networks on an on-demand, point-to-point basis, but does not include services providing video programming prescheduled by the programming provider. [47 USC 522, 47 USC 653]

interconnected (see interconnection)

interconnected service [1]
Service that is interconnected with the public switched network (as such terms are defined by regulation by the Commission) or service for which a request for interconnection is pending pursuant to subsection (c)(1)(B). [47 USC 332]

interconnected service [2]
A service (a) that is interconnected with the public switched network, or interconnected with the public switched network through an interconnected service provider, that gives subscribers the capability to communicate to or receive the communication from all other users on the public switched network; or (b) for which a request for such interconnection is pending pursuant to section 332(c)(1)(B) of the Communications Act, 47 U.S.C. 332(c)(1)(B). A mobile service offers interconnected service even if the service allows subscribers to access the public switched network only during specified hours of the day, or if the service provides general access to points on the public switched network but also restricts access in certain limited ways. Interconnected service does not include any interface between a licensee's facilities and the public switched network exclusively for a licensee's internal control purposes. [47 CFR 20.3]

interconnection [1]

The use of microwave equipment, boosters, translators, repeaters, communication space satellites, or other apparatus or equipment for the transmission and distribution of television or radio programs to public telecommunications entities. [47 USC 397]

interconnection (interconnected) [2]

Direct or indirect connection through automatic or manual means (by wire, microwave, or other technologies such as store and forward) to permit the transmission or reception of messages or signals to or from points in the public switched network. [47 CFR 20.3]

interconnection [3]

Connection through automatic or manual means of private land mobile radio stations with the facilities of the public switched telephone network to permit the transmission of messages or signals between points in the wireline or radio network of a public telephone company and persons served by private land mobile radio stations. Wireline or radio circuits or links furnished by common carriers, which are used by licensees or other authorized persons for transmitter control (including dial-up transmitter control circuits) or as an integral part of an authorized, private, internal system of communication or as an integral part of dispatch point circuits in a private land mobile radio station are not considered to be interconnection for purposes of this rule part. [47 CFR 90.7]

interconnection system

Any system of interconnection facilities used for distribution of programs to public telecommunications entities. [47 USC 397]

interexchange (interexchange category)

Includes services or facilities provided as an integral part of interstate or foreign telecommunications that is not described as "access service" for purposes of this part. [47 CFR 69.2]

interexchange channel

A circuit which is included in the interexchange transmission equipment. [47 CFR Pt. 36, App.]

interexchange transmission equipment

The combination of (a) interexchange cable and wire facilities, (b) interexchange circuit equipment and, (c) associated land and buildings. [47 CFR Pt. 36, App.]

interface

The point of interconnection between two distinct but adjacent communications systems having different functions. The interface in the communication-satellite service is that point where communications terminal equipment of the terrestrial common carriers or other authorized entities interconnects with the terminal equipment of the communication-satellite earth station complex. The interface in the communication-satellite service shall be located at the earth station site, or if this is impracticable, as close thereto as possible. [47 CFR 25.103]

interference

The effect of unwanted energy due to one or a combination of emissions, radiations, or inductions upon reception in a radiocommunication system, manifested by any performance degradation, misinterpretation, or loss of information which could be extracted in the absence of such unwanted energy. (RR) [47 CFR 2.1]

interfering contour

The locus of points surrounding a transmitter where the predicted median field strength of the signal from that transmitter is the maximum field strength that is not considered to cause interference at the service contour of another transmitter. [47 CFR 22.99]

interlaced scanning

A scanning process in which successively scanned lines are spaced an integral number of line widths, and in which the adjacent lines are scanned during successive cycles of the field frequency. [47 CFR 73.681]

InterLATA service

Telecommunications between a point located in a local access and transport area and a point located outside such area. [47 USC 153]

interlocal trunk

A circuit between two local central office units, either manual or dial. Interlocal trunks may be used for either exchange or toll traffic or both. [47 CFR Pt. 36, App.]

interlocking director

A person who performs the duties of "officer of director" in more than one carrier subject to the Communications Act of 1934, as amended. [47 CFR 62.2]

intermittent service area

The area receiving service from the groundwave of a broadcast station but beyond the primary service area and subject to some interference and fading. [47 CFR 73.14]

internal system

An internal system of communication is one in which all messages are transmitted between the fixed operating positions located on premises controlled by the licensee and the associated mobile stations or paging receivers of the licensee. (See Subpart O). [47 CFR 90.7]

international broadcasting stations

A broadcasting station employing frequencies allocated to the broadcasting service between 5,950 and 26,100 kHz, the transmission of which are intended to be received directly by the general public in foreign countries. (A station may be authorized more than one transmitter.) There are both government and non-government international broadcasting stations; only the latter are licensed by the Commission and are subject to the rules of this subpart. [47 CFR 73.701]

international fixed public control service

A fixed service carried on for the purpose of communicating between transmitting stations, receiving stations, message centers or control points in the international fixed public radiocommunication service. [47 CFR 23.1]

international fixed public radiocommunication service

A fixed service, the stations of which are open to public correspondence and which, in general, is intended to provide radiocommunication between any one of the contiguous 48 states (including the District of Columbia) and the state of Alaska, or the state of Hawaii, or any U.S. possession or any foreign point; or between any U.S. possession and any other point; or between the state of Alaska and any other point; or between the state of Hawaii and any other point. In addition, radiocommunications within the contiguous 48 states (including the District of Columbia) in connection with the relaying of international traffic between stations which provide the above service, are also deemed to be the international fixed public radiocommunications service; provided, however, that communications solely between Alaska, or any one of the contiguous 48 states (including the District of Columbia), and either Canada or Mexico are not deemed to be in the international fixed public radiocommunication service when such radiocommunications are transmitted on frequencies above 72 MHZ. [47 CFR 23.1]

International Morse Code

A dot-dash code as defined in International Telegraph and Telephone Consultative Committee (CCITT) Recommendation F.1 (1984), Division B, I. Morse code. [47 97.3, 13.3]

Internet

The international computer network of both federal and non-federal interoperable packet switched data networks. [47 USC 230]

internodal link [1]

Two point-to-point microwave radio stations used to provide two-way communications between Digital Termination Nodal Stations or to interconnect Digital Termination Systems to other communications media. [47 CFR 21.2]

internodal link [2]

A point-to-point communications link used to provide communications between Nodal Stations or to interconnect Nodal Stations to other communications media. [47 CFR 94.3]

interoffice transmitter

A fixed transmitter in the Rural Radiotelephone Service that communicates with other interoffice transmitters for the purpose of interconnecting rural central offices. [47 CFR 22.99]

inter-satellite service

A radiocommunication service providing links between artificial earth satellites. (RR) [47 CFR 2.1, 25.201]

interstate communication (interstate transmission)

Communication or transmission (A) from any state, territory, or possession of the United States (other than the [Philippine Islands and] and Canal Zone), or the District of Columbia, to any other state, Territory, or possession of the United States (other than [the Philippine Islands, and] the Canal Zone), or the District of Columbia, (B) from or to the United States to or from [the Philippine Islands] or the Canal Zone, insofar as such communication or transmission takes place within the United States, or (C) between points within the United States but through a foreign country; but shall not, with respect to the provisions of Title II of this act (other than Section 223 thereof), include wire or radio communication between points in the same state, Territory, or possession of the United States, or the District of Columbia, through any place outside thereof, if such communication is regulated by a state commission. [47 USC 153]

intertoll circuits

Circuits between toll centers and circuits between a toll center and a tandem system in a different toll center area. [47 CFR Pt. 36, App.]

intrasystems

Assets consisting of (1) PBX and Key System Common Equipment (a switchboard or switching equipment shared by all stations); (2) associated CPE station equipment (usually telephone or Key Telephone Systems); and (3) intrasystem wiring (all cable or wiring and associated components which connect the common equipment and the station equipment, located on the customer's side of the demarcation point).

An intrasystem does not include property, plant or equipment which are not solely dedicated to its operation. [47 CFR 32.9000]

IOC

Initial operating capability [47 CFR 64.1604]

ionospheric scatter

The propagation of radio waves by scattering as a result of irregularities or discontinuities in the ionization of the ionosphere. (RR) [47 CFR 2.1]

IOT

Information origination/termination equipment [47 CFR 69.303]

IRAC

Interdepartment Radio Advisory Committee [47 CFR 25.142]

IRC

Incrementally related carriers [47 CFR 76.612]

IRE standard scale

A linear scale for measuring, in IRE units, the relative amplitudes of the components of a television signal from a zero reference at blanking level, with picture information falling in the positive, and synchronizing information in the negative domain. [47 CFR 73.681]

ISM

Industrial, scientific and medical [47 CFR 0.314]

ISM frequency

A frequency assigned by this part for the use of ISM equipment. A specified tolerance is associated with each ISM frequency. See §18.301. [47 CFR 18.107]

isochronous devices

Devices that transmit at a regular interval, typified by time-division voice systems. [47 CFR 15.303]

issuing carrier

A carrier subject to the Act that publishes and files a tariff or tariffs with the Commission. [47 CFR 61.3]

ITFS

Instructional television fixed service [47 CFR 21.902]

ITFS response station
     A fixed station operated at an authorized location to provide communication by voice and/or data signals to an associated instructional television fixed station. [47 CFR 74.901]

itinerant operation
     Operation of a radio station at unspecified locations for varying periods of time. [47 CFR 90.7]

ITT (see text telephone)

ITU
     International Telecommunication Union [47 CFR 0.261]

IVDS
     Interactive video and data service [47 CFR 95.801]

IXC
     Interexchange carrier [47 CFR 64.1100]

J K

JTIDS

Joint tactical information distribution systems [47 CFR 87.479]

kHz

Kilohertz [47 CFR 2.106]

kinship affiliation

Immediate family members will be presumed to own or control or have the power to control interests owned or controlled by other immediate family members. In this context "immediate family member" means father, mother, husband, wife, son, daughter, brother, sister, father- or mother-in-law, son- or daughter-in-law, brother- or sister-in-law, step-father, or -mother, step-brother, or -sister, step-son, or -daughter, half brother or sister. This presumption may be rebutted by showing that (A) the family members are estranged, (B) the family ties are remote, or (C) the family members are not closely involved with each other in business matters. [47 CFR 24.720, 90.814]

kit

Any number of electronic parts, usually provided with a schematic diagram or printed circuit board, which, when assembled in accordance with instructions, results in a device subject to the regulations of this part, even if additional parts of any type are required to complete assembly. [47 CFR 15.3]

KTS

Key telephone systems [47 CFR 68.316]

L

land earth station

An earth station in the fixed-satellite service or, in some cases, in the mobile-satellite service, located at a specified fixed point or within a specified area on land to provide a feeder link for the mobile-satellite service. (RR) [47 CFR 2.1, 25.201]

land mobile earth station

A mobile earth station in the land mobile-satellite service capable of surface movement within the geographical limits of a country or continent. (RR) [47 CFR 2.1, 24.5, 25.201]

land mobile radio system

A regularly interacting group of base, mobile and associated control and fixed relay stations intended to provide land mobile radio communications service over a single area of operation. [47 CFR 90.7]

land mobile-satellite service

A mobile-satellite service in which mobile earth stations are located on land. (RR) [47 CFR 2.1]

land mobile service

A mobile service between base stations and land mobile stations, or between land mobile stations. (RR) [47 CFR 2.1, 24.5, 26.4, 90.7]

land station [1]

A station, other than a mobile station, used for radio communication with mobile stations. [47 USC 153]

land station [2]

A station in the mobile service not intended to be used while in motion. (As used in this part, the term may be used to describe a base, control, fixed, operational fixed or fixed relay station, or any such station authorized to operate in the "temporary" mode.) [47 CFR 2.1, 5.4, 24.5, 26.4, 87.5, 90.7]

land station [3]

A unit which transmits only from (1) an exact point as shown on the license; or (2) an unspecified point within an *operating area* (an area within a circle centered on point chosen by the applicant) as shown on the license, for a *temporary period* (one year or less). [47 CFR 95.25]

landing area [1]

As defined by Title I, section I (22) of the Civil Aeronautics Act of 1938, as amended, landing area means any locality, either of land or water, including airdromes and intermediate landing fields, which is used, or intended to be used, for the landing and take-off of aircraft, whether or not facilities are provided for the shelter, servicing, or repair of aircraft, or for receiving or discharging passengers or cargo. [47 CFR 5.4]

landing area [2]

Any locality, either of land or water, including airports and intermediate landing fields, which is used, or approved for use for the landing and takeoff of aircraft, whether or not facilities are provided for the shelter, servicing, or repair of aircraft, or for receiving or discharging passengers or cargo. [47 CFR 21.2]

last radio stage

The radio-frequency power amplifier stage which supplies power to the antenna. [47 CFR 73.14]

LDS

Local distribution service [47 CFR 78.5]

LEC

Local exchange carrier [47 CFR 64.1100]

LEC joint use card

A calling card bearing an account number assigned by a local exchange carrier, used for the services of the local exchange carrier and a designated interexchange carrier, and validated by access to data maintained by the local exchange carrier. [47 CFR 64.1201]

left (or right) signal

The electrical output of a microphone or combination of microphones placed so as to convey the intensity, time, and location of sounds originated predominantly to the listener's left (or right) of the center of the performing area. [47 CFR 73.14, 73.310]

left (or right) stereophonic channel [1]

The left (or right) signal as electrically reproduced in reception of AM stereophonic broadcasts. [47 CFR 73.14]

left (or right) stereophonic channel [2]

The left (or right) signal as electrically reproduced in reception of FM stereophonic broadcasts. [47 CFR 73.310]

left-hand (or anti-clockwise) polarized wave

An elliptically or circularly-polarized wave, in fixed plane, normal to the direction of propagation, whilst looking in the direction of propagation, rotates with time in a left hand or anti-clockwise direction. (RR) [47 CFR 2.1]

legally qualified candidate

Any person who (i) has publicly announced his or her intention to run for nomination or office; (ii) is qualified under the applicable local, state or federal law to hold the office for which he or she is a candidate; and (iii) has met the qualifications set forth in either paragraphs (q) (2), (3) or (4) of this section. [47 CFR 76.5]

level I contributors

Telephone companies that are not association Common Line tariff participants, file their own Common Line tariffs effective April 1, 1989, and had a lower than average Common Line revenue requirement per minute of use in 1988 and thus were net contributors (*i.e.*, had a negative net balance) to the association common Line pool in 1988. [47 CFR 69.2]

level II contributors

A telephone company or group of affiliated telephone companies with fewer than 300,000 access lines and less than $150 million in annual operating revenues that is not an association Common line tariff participant, that files its own Common Line tariff effective July 1, 1990, and that had a lower than average Common Line revenue requirement per minute of use in 1988 and thus was a net receiver (*e.g.*, had a positive net balance) from the association Common Line pool in 1988. [47 CFR 69.2]

level I receivers

Telephone companies that are not association Common Line tariff participants, file their own common Line tariffs effective April 1, 1989, and had a higher than average Common Line revenue requirement per minute of use in 1988 and thus were net receivers (*i.e.*, had a positive net balance) from the association Common Line Pool in 1988. [47 CFR 69.2]

level II receivers

A telephone company or group of affiliated telephone companies with fewer than 300,000 access lines and less than $150 million in annual operating revenues that is not an association Common Line tariff participant, that files its own Common Line tariff effective July 1, 1990, and that had a higher than average Common Line revenue requirement per minute of use in 1988 and thus was a net receiver (*i.e.*, had a positive net balance) from the association Common Line pool in 1988. [47 CFR 69.2]

library

A library eligible for participation in state-based plans for funds under title III of the Library Services and Construction Act (20 USC 355e et seq.). [47 USC 223]

license (see station license)

licensee

The holder of a radio station license granted or continued in force under authority of this Act. [47 USC 153]

licensee (station licensee)

When used with respect to a community antenna television system, the operator of such system. [47 USC 315]

line (trunk)

Includes, but is not limited to, transmission media such as radio, satellite, wire, cable and fiber optic cable means of transmission. [47 CFR 69.2]

line A

An imaginary line within the U.S., approximately paralleling the U.S.-Canadian border, north of which Commission coordination with Canadian authorities in the assignment of frequencies is generally required. It begins at Aberdeen, Washington, running by great circle arc to the intersection of 48° N., 120° W., then along parallel 48° N., to the intersection of 95° W., thence by great circle arc through the southern most point of Duluth, Minn., thence by great circle arc to 45° N., 85° W., thence southward along meridian 85° W., to its intersection with parallel 41° N., thence along parallel 41° N. to its intersection with meridian . . . 82° W., thence by great circle arc through the southernmost point of Bangor, Maine, thence by great circle arc through the southernmost point of Searsport, Maine, at which point it terminates. [47 CFR 2.1, 90.7, 97.3]

line B

Begins at Torino, B.C., running by great circle arc to the intersection of 50° N., 125° W., thence along parallel 50° N., to the intersection of 90° W., thence by great circle arc to the intersection 45° N., 79° 30' W., thence by great circle arc through the northernmost point of Drummondville, Quebec (Lat. 45° 52' N., Long 72° 30' W.), thence by great circle arc to 48° 30' N., 70° W., thence by great circle arc through the northernmost point of Combellton, N.B., thence by great circle are through the northernmost point of Liverpool, N.S., at which point it terminates. (FCC) [47 CFR 2.1]

line C

An imaginary line in Alaska approximately paralleling the border with Canada, East of which Commission coordination with Canadian authorities in the assignment of frequencies is generally required. It begins at the intersection of 70° N., 144° W., thence by great circle arc to the intersection of 60° N., 143° W., thence by great circle arc so as to include all the Alaskan Panhandle. [47 CFR 2.1, 90.7]

line D

Begins at the intersection of 70° N., 138° W., thence by great circle arc to the intersection of 61°20' N., 139° W. (Burwash Landing), thence by great circle arc to the intersection of 60°45' N. 135° W., thence by great circle arc to the intersection 56° N., 128° W., thence south along 128° meridian to Lat. 55°N. Thence by great circle arc to the intersection of 54° N., 130° W., thence by great circle arc to Port Clements, thence to the Pacific Ocean where it ends. (FCC) [47 CFR 2.1]

LISN

Line impedance stabilization network [47 CFR 18.307]

listening or viewing public

Those members of the public who, with the aid of radio receiving sets, listen to or view programs broadcast by radio stations. [47 USC 508]

LMS

Location and monitoring service [47 CFR 2.106]

local access and transport area (LATA)

A contiguous geographic area (A) established before the date of enactment of the Telecommunications Act of 1996 by a Bell operating company such that no exchange area includes points within more than 1 metropolitan statistical area, consolidated metropolitan statistical area, or state, except as expressly permitted under the AT&T Consent Decree; or (B) established or modified by a Bell operating company after such date of enactment and approved by the Commission. [47 USC 153]

local area data channel (LADC) leads

Terminal equipment leads at the interface used to transmit and/or receive signals which may require greater-than-voiceband frequency spectrum over private line metallic channels designated Local Area Data Channels (LADC). These leads should be treated as "telephone connections" as defined in this section or as tip and ring connections where the term "telephone connection" is not used. [47 CFR 68.3]

local area data channel simulator circuit

A circuit for connection in lieu of a Local Area Data Channel to provide the appropriate impedance for signal power tests. The schematic of

Figure 68.3(k) is illustrative of the type of circuit that will be required over the given frequency ranges. When used, the simulator shall be operated over the appropriate range of loop resistance for the equipment under test, under all voltages and polarities that the terminal under test and a connected companion unit are capable of providing. [47 CFR 68.3]

local channel
The portion of a private line circuit which is included in the exchange transmission plant. · However, common usage of this term usually excludes information origination/termination equipment. [47 CFR Pt. 36, App.]

local commercial television station
Any full power television broadcast station, other than a qualified noncommercial educational television station within the meaning of Section 615(1)(1), licensed and operating on a channel regularly assigned to its community by the Commission that, with respect to particular cable system, is within the same television market as the cable system.
The term shall not include (i) low power television stations, television translator stations, and passive repeaters which operate pursuant to Part 74 of Title 47, Code of Federal Regulations, or any successor regulations thereto; (ii) a television broadcast station that would be considered a distant signal under Section 111 of Title 17, United States Code, if such station does not agree to indemnify the cable operator for any increased copyright liability resulting from carriage on the cable system; or (iii) a television broadcast station that does not deliver to the principal headend of a cable system either a signal level of -45 dBm for UHF signals or -49 dBm for VHF signals at the input terminals of the signal processing equipment, if such station does not agree to be responsible for the costs of delivering to the cable system a signal of good quality or a baseband video signal. [47 USC 534, 47 CFR 76.55]

local control
The use of a control operator who directly manipulates the operating adjustments in the station to achieve compliance with the FCC Rules. [47 CFR 97.3]

local distribution service (LDS) station
A fixed CARS station used within a cable television system or systems for the transmission of television signals and related audio signals, signals of standard and FM broadcast stations, signals of instructional

television fixed stations, and cablecasting from a local transmission point to one or more receiving points, from which the communications are distributed to the public. LDS stations may also engage in repeated operation. [47 CFR 78.5]

local exchange carrier [1]

Any person that is engaged in the provision of telephone exchange service or exchange access. Such term does not include a person insofar as such person is engaged in the provision of a commercial mobile service under Section 332(c), except to the extent that the Commission finds that such service should be included in the definition of such term. [47 USC 153]

local exchange carrier [2]

A telephone company that provides telephone exchange service as defined in section 3(r) of the Act. [47 CFR 61.3]

local office

A central office serving primarily as a place of termination for subscriber lines and for providing telephone service to the subscribers on these lines. [47 CFR Pt. 36, App.]

local origination

Program origination of the parameters of the program source signal as it reaches the transmitter site, are under the control of the low power TV station licensee. Transmission of TV program signals generated at the transmitter site constitutes local origination. Local origination also includes transmission of programs reaching the transmitter site via TV STL stations, but does not include transmission of signals obtained from either terrestrial or satellite microwave feeds or low power TV stations. [47 CFR 74.701]

local taxing jurisdiction

Any municipality, city, county, township, parish, transportation district, or assessment jurisdiction, or any other local jurisdiction in the territorial jurisdiction of the United States with the authority to impose a tax or fee, but does not include a state. [47 USC 602]

local television transmission service

A domestic public radio communication service for the transmission of television material and related communications. [47 CFR 21.2]

localizer station

A radionavigation land station which provides horizontal guidance to aircraft with respect to a runway center line. [47 CFR 87.5]

location and monitoring service (LMS)

The use of non-voice signaling methods to locate or monitor mobile radio units. LMS systems may transmit and receive voice and nonvoice status and instructional information related to such units. [47 CFR 90.7]

long haul system

A microwave system licensed under this part in which the longest radio circuit of tandem radio paths exceeds 402 km (250 miles). [47 CFR 94.3]

long term support (LTS)

Funds provided by telephone companies that are not association Common Line tariff participants to association Common Line tariff participants. LTS enables association Common Line tariff participants to charge a Common Line (CL) rate equivalent to the CL rate that would result if all telephone companies participated in the association Common Line tariff. [47 CFR 69.2]

longitudinal voltage

One half of the vector sum of the potential difference between the tip connection and earth ground, and the ring connection and earth ground for the tip, ring pair of 2-wire and 4-wire connections; and, additionally for 4-wire telephone connections, one half of the vector sum of the potential difference between the tip 1 connection and earth ground and the ring 1 connection and earth ground for the tip 1, ring 1 pair (where tip 1 and ring 1 are the receive pair). [47 CFR 68.3]

loop

A pair of wires, or its equivalent, between a customer's station and the central office from which the statio is served. [47 CFR Pt. 36, App.]

loop simulator circuit

A circuit that simulates the network side of a 2-wire or 4-wire telephone connection during testing. The required circuit schematics are shown in Figure 68.3(a) for 2-wire loop or ground start circuits, Figure 68.3(b) for 2-wire reverse battery circuits, Figure 68.3(c) for 4-wire loop or ground start circuits, Figure 68.3(d) for 4-wire reverse battery circuits, and Figure 68.3(j) for voice band metallic channels. Figure 68.3(i) is an alternative termination for use in the 2-wire loop simulator circuits. Other implementations may be used provided that the same dc voltage and current characteristics and ac impedance characteristics will be presented to the equipment under test as are presented in the illustrative schematic diagrams. When used, the simulator shall be operated over the entire range of loop resistance as indicated in the figures, and with the indicated polarities and voltage limits. Whenever loop current is changed, sufficient time shall be allocated for the current to reach a steady-state condition before continuing testing. [47 CFR 68.3]

LORAN
        A radio-navigation system. [47 CFR 2.106]

low power auxiliary station
        An auxiliary station authorized and operated pursuant to the provisions set forth in this subpart. Devices authorized as low power auxiliary stations are intended to transmit over distances of approximately 100 meters for uses such as wireless microphones, cue and control communications, and synchronization of TV camera signals. [47 CFR 74.801]

low power TV station
        A station authorized under the provisions of this subpart that may retransmit the programs and signals of a TV broadcast station and that may originate programming in any amount greater than 30 seconds per hour and/or operates a subscription service (See §73.641 of part 73 of this chapter.) [47 CFR 74.701]

LP
        Local primary [47 CFR 11.18]

LPTV
        Low power television [47 CFR 11.11, 76.7]

LTS (see long term support)

luminance

Luminous flux emitted, reflected, or transmitted per unit solid angle per unit projected area of the source. [47 CFR 73.681]

M

m
> Meters [47 CFR 97.3]

magnetic resonance equipment
> A category of ISM equipment in which RF energy is used to create images and data representing spatially resolved density of transient atomic resources within an object. [47 CFR 18.107]

main channel [1]
> The band of audio frequencies from 50 to 10,000 Hz which amplitude modulates the carrier. [47 CFR 73.14]

main channel [2]
> The band of frequencies from 50 to 15,000 Hertz which frequency modulate the main aural carrier. [47 CFR 73.310, 73.681]

main channel [3]
> That portion of each authorized channel used for the transmission of visual and aural information as set forth in §73.682 of this chapter and §74.938 of this subpart.[47 CFR 74.901]

major airports
> Those airports described by the Federal Aviation Administration as large or medium hubs. The member agencies of the National Communications System (NCS) will determine which of their locations are "major military installations" and "key government facilities." [47 CFR 63.100]

major television market

The specified zone of a commercial television station licensed to a community listed in §76.51, or a combination of such specified zones where more than one community is listed. [47 CFR 76.5]

major trading areas

Service areas based on the Rand McNally *1992 Commercial Atlas &Marketing Guide*, 123rd Edition at pages 38-39, with the following exceptions and additions: (a) Alaska is separated from the Seattle MTA and is licensed separately. (b) Guam and the Northern Mariana Islands are licensed as a single MTA-like area. (c) Puerto Rico and the United States Virgin Islands are licensed as a single MTA-like area. (d) American Samoa is licensed as a single MTA-like area. [47 CFR 90.7]

make-busy leads

Terminal equipment leads at the network interface designated MB and MB1. The MB lead is connected by the terminal equipment to the MB1 lead when the corresponding telephone line is to be placed in an unavailable or artificially busy condition. [47 CFR 68.3]

March season

That portion of any year commencing 0100 g.m.t. on the first Sunday in March and ending at 0100 g.m.t. on the first Sunday in May. [47 CFR 73.701]

marine utility station

A station in the maritime mobile service consisting of one or more handheld radiotelephone units licensed under a single authorization. Each unit is capable of operation while being hand-carried by an individual. The station operates under the rules applicable to ship stations when the unit is aboard a vessel, and under the rules applicable to private coast stations when the unit is on land. [47 CFR 80.5]

MARISAT

A satellite for marine use. [47 CFR 80.51]

maritime control communications

Communications between private coast and ship stations or between ship stations licensed to a state or local governmental entity, which relate directly to the control of boating activities or assistance to ships. [47 CFR 80.5]

maritime mobile repeater station

A land station at a fixed location established for the automatic retransmission of signals to extend the range of communication of ship and coast stations. [47 CFR 80.5]

maritime mobile-satellite service

A mobile-satellite service in which mobile earth stations are located on board ships; survival craft stations and emergency position-indicating radiobeacon stations may also participate in this service. (RR) [47 CFR 2.1, 80.5]

maritime mobile service

A mobile service between coast stations and ship stations, or between ship stations, or between associated on-board communication stations; survival craft stations and emergency position-indicating radiobeacon stations may also participate in this service. (RR) [47 CFR 2.1, 80.5]

maritime mobile service identities

An international system for the identification of radio stations in the maritime mobile service. The system is comprised of a series of nine digits which are transmitted over the radio path to uniquely identify ship stations, ship earth stations, coast stations, coast earth stations and groups of stations. [47 CFR 80.5]

maritime radiodetermination service

A maritime radiocommunication service for determining the position, velocity, and/or other characteristics of an object, or the obtaining of information relating to these parameters, by the propagation properties of radio waves. [47 CFR 80.5]

maritime radionavigation-satellite service

A radionavigation-satellite service in which earth stations are located on board ships. (RR) [47 CFR 2.1]

maritime radionavigation service
A radionavigation service intended for the benefit and for the safe operation of ships. (RR) [47 CFR 2.1]

maritime support station
A station on land used in support of the maritime services to train personnel and to demonstrate, test and maintain equipment. [47 CFR 80.5]

marker beacon
A transmitter in the aeronautical radionavigation service which radiates vertically a distinctive pattern for providing position information to aircraft. (RR) [47 CFR 2.1]

marker beacon station
A radionavigation land station in the aeronautical radionavigation service which employs a marker beacon. A marker beacon is a transmitter which radiates vertically a distinctive pattern for providing position information to aircraft. [47 CFR 87.5]

marketing
Marketing shall include sale or lease, offer for sale or lease, advertising for sale or lease, the import or shipment or other distribution for the purpose of sale or lease or offer for sale or lease. See subpart I of part 2 of this chapter. [47 CFR 18.107]

master antenna
The television interface devices employed for central distribution of television or other video signals within a building. Such TV interface devices must be designed to (i) distribute multiple television signals at the same time; (ii) distribute such signals by cable to outlets or TV receivers in multiple rooms in the building in which the TV interface devices are installed; and (iii) distribute all over-the-air or cable signals. Note: Cable-ready video cassette recorders continue to be subject to the provisions for general TV interface devices. [47 CFR 15.115]

master station
A station in a multiple address radio system that controls, activates, or interrogates four or more remote stations. Master stations performing such functions may also receive transmissions from remote stations. [47 CFR 94.3]

material terms

Those factors which define the operation of the contest and which affect participation therein. Although the material terms may vary widely depending upon the exact nature of the contest, they will generally include: how to enter or participate; eligibility restrictions; entry deadline dates; whether prizes can be won; when prizes can be won; and value of prizes; basis for valuation of prizes; time and means of selection of winners; and/or tie-breaking procedures.
[47 CFR 73.1225]

maximum percentage of modulation

The greatest percentage of modulation that may be obtained by a transmitter without producing in its output, harmonics of the modulating frequency in excess of those permitted by these regulations. (See §73.1570)
[47 CFR 73.14]

maximum rated carrier power

The maximum power at which the transmitter can be operated satisfactorily and is determined by the design of the transmitter and the type and number of vacuum tubes or other amplifier devices used in the last radio stage. [47 CFR 73.14]

maximum usable frequency (MUF)

The highest frequency which is returned by ionospheric radio propagation to the surface of the earth for a particular path and time of day for 50 percent of the days of the reference month. [47 CFR 73.701]

May season

That portion of any year commencing at 0100 G.M.T. on the first Sunday in May and ending at 0100 G.M.T. on the first Sunday in September. [47 CFR 73.701]

MB/MB1

Connections to leads implementing a make-busy feature where required. The MB lead is shorted by the terminal equipment to the MB1 lead when the corresponding telephone line is to be placed in an unavailable, or artificially busy condition. [47 CFR 68.502]

McCaw Consent Decree

The proposed Consent decree filed on July 15, 1994, in the antitrust action styled United States v. AT&T Corp. and McCaw Cellular Communications, Inc. Civil Action No. 94-01555, in the United States District Court for the District of Columbia. Such term includes any stipulation that the parties will abide by the terms of such proposed consent decree until it is entered and any order entering such proposed consent decree. [47 USC 601]

MCD

Minor civil division [47 CFR 73.525]

MCW

Tone-modulated international Morse code telegraphy emissions having designators with A, C, D, F, G, H or R as the first symbol; 2 as the second symbol; A or B as the third symbol. [47 CFR 97.3]

MDS

Multipoint distribution service [47 CFR 1.1622]

MDS service areas

MDS service areas are regional basic trading areas (BTAs) [47 CFR 21.924]

mean power [1]

The power at the output terminals of a transmitter during normal operation, averaged over a time sufficiently long compared with the period of the lowest frequency encountered in the modulation. A time of 1/10 second during which the mean power is greatest will be selected normally. [47 CFR 74.401]

mean power [2]

TP averaged over at least 30 cycles of the lowest modulating frequency, typically 0.1 seconds at maximum power. [47 CFR 95.669]

mean power (of a radio transmitter) [1]

The average power supplied to the antenna transmission line by a transmitter during an interval of time sufficiently long compared with the lowest frequency encountered in the modulation taken under normal operating conditions. (RR) [47 CFR 2.1, 87.5]

mean power of radio transmitter [2]

The power supplied to the antenna during normal operation, averaged over a time sufficiently long compared to the prior corresponding to the lowest frequency encountered in actual modulation. [47 CFR 5.4]

measurement axis

Parallel to the reference axis but may be displaced from that axis, by a maximum of 10 mm (see Fig 1). Within this constraint, the measurement axis may be located where the axial and radial field intensity measurements, are optimum with regard to the requirements. In a handset with a centered receiver and a circularly symmetrical magnetic field, the measurement axis and the reference axis would coincide. [47 CFR 68.316]

media of mass communications

Television, radio, cable television, multi point distribution service, direct broadcast satellite service, and other services, the licensed facilities of which may be substantially devoted toward providing programming or other information services within the editorial control of the licensee. [47 USC 309]

medical diathermy equipment

A category of ISM equipment used for therapeutic purposes, not including surgical diathermy apparatus designed for intermittent operation with low power. [47 CFR 18.107]

medium frequency (MF)

The frequency range 300-3000 kHz. [47 CFR 97.3]

meeting [1]

The deliberations of at least the number of members of a governing or advisory body, or any committee thereof, required to take action on behalf of such body or committee were such deliberations determine or result in the joint conduct or disposition of the governing of advisory body's business, or the committee's business, as the case may be, but only to the extent that such deliberations relate to public broadcasting. [47 USC 397]

meeting [2]

The deliberations among a quorum of the commission, a Board of Commissioners, or a quorum of a committee of Commissioners, where such deliberations determine or result in the joint conduct of disposition of official agency business, except that the term does not include deliberations to decide whether a meeting should be open or closed. (The term includes conference telephone calls, but does not include the separate consideration of Commission business by Commissioners.) For purposes of this subpart each item on the agenda of a meeting is considered a meeting or a portion of a meeting. [47 CFR 0.601]

members of minority groups [1]
Includes individuals of African American, Hispanic-surnamed, American Eskimo, Aleut, American Indiana and Asian American extraction. [47 CFR 24.230]

members of minority groups [2]
Includes Blacks, Hispanics, American Indians, Alaskan Natives, Asians, and Pacific Islanders. [47 CFR 24.720, 90.814]

MERP
Maximum effective radiated power [47 CFR 74.1235]

message
A completed call, *i.e.*, a communication in which a conversation or exchange of information took place between the calling and called parties. [47 CFR Pt. 36, App.]

message center
The point at which messages from members of the public are accepted by the carrier for transmission to the addressee. [47 CFR 21.2]

message forwarding system
A group of amateur stations participating in a voluntary, cooperative, interactive arrangement where communications are sent from the control operator of an originating station to the control operator of one or more destination stations by one or more forwarding stations. [47 CFR 97.3]

message register leads
Terminal equipment leads at the interface used solely for receiving dc message register (MR) pulses from a central office at a PBX so that message unit information normally recorded at the central office only is also recorded at the PBX. Signaling on the channel is by the applications of battery and

open conditions applied at the central office. No ac signaling is applied either by the PBX or by the central office. One or more pairs of MR leads, each designated T (MR) and R (MR) may appear at an interface depending on the number of PBX-CO trunks (one MR cannel per PBX-CO trunk). However, unless otherwise stated, these leads at the interface should be treated at telephone connections as defined in paragraph (x) of this section or as tip and ring where the term "telephone connection" is not used. [47 CFR 68.3]

message register signaling channel simulator

A circuit that simulates a telephone line (2-wire or single conductor) and a central office message register battery feed circuit used to convey message register information from the central office to a PBX. The schematic of Figure 68.3(h) is illustrative of the type of circuit that will be required; alternative implementation may be used provided that the same dc voltage and current characteristics and ac impedance characteristics will be presented to the message register equipment under test. When used, the simulator circuit shall be operated over the entire range of resistance and voltage values indicated in Figure 68.3(h). Whenever dc current is changed, sufficient time shall be allocated for the current to reach a steady-state condition before continuing the test. [47 CFR 68.3]

message service or message toll service

Switched service furnished to the general public (as distinguished from private line service). Except as otherwise provided, this includes exchange switched services and all switched services provided by interexchange carriers and completed by a local telephone company's access services, *e.g.*, MTS, WATS, Execunet, open-end FX and CCSA/ONALs. [47 CFR Pt. 36, App.]

message units

Unit of measurement used for charging for measured message telephone exchange traffic within a specified area. [47 CFR Pt. 36, App.]

metallic voltage

The potential difference between the tip and ring connections for the tip, ring pair of 2-wire and 4-wire connections and additionally for 4-wire telephone connections, between the tip 1 and ring 1 connections with provisions internal to the equipment for establishing transmission paths among two or more telephone connections. [47 CFR 68.3]

meteor burst communications

Communications by the propagation of radio signals reflected off ionized meteor trails. [47 CFR 90.7]

meteor burst propagation mode

A long distance VHF radio communication path occurring as a result of the refraction of electromagnetic waves by ionized meteor trails. [47 CFR 22.99]

meteorological aids service

A radiocommunication service used for meteorological, including hydrological, observation and exploration.(RR) [47 CFR 2.1]

meteorological-satellite service

An earth exploration-satellite service for meteorological purposes.(RR) [47 CFR 2.1]

metropolitan service area

The area around and including a relatively large city and in which substantially all of the message telephone traffic between the city and the suburban points within the area is classified as an exchange in one or both directions. [47 CFR Pt. 36, App.]

MF

Medium frequency [47 CFR 97.3]

MHZ

Megahertz [47 CFR 1.924]

microwave

For the purposes of this part, frequencies from 928-929 MHZ and those above 952 MHZ. [47 CFR 94.3]

microwave frequencies

As used in this part, this term refers to frequencies of 890 MHZ and above. [47 CFR 21.2]

microwave landing system
An instrument landing system operating in the microwave spectrum that provides lateral and vertical guidance to aircraft having compatible avionics equipment. [47 CFR 87.5]

minimum point of entry
Either the closest practicable point to where the wiring crosses a property line or the closest practicable point to where the wiring enters a multiunit building or buildings. The telephone company's reasonable and nondiscriminatory standard operating practices shall determine which shall apply. The telephone company is not precluded from establishing reasonable classifications of multiunit premises for purposes of determining which shall apply. Multiunit premises include, but are not limited to, residential, commercial, shopping center and campus situations. [47 CFR 68.3]

minor items
As applied to depreciable telecommunications plant, means any part or element of such plant, which when removed, (with or without replacement) does not initiate retirement accounting. [47 CFR 32.9000]

minority
Black, Hispanic, American Indian, Alaska Native, Asian and Pacific Islander. [47 CFR 73.3555]

minority-controlled
More than 50 percent owned by one or more members of a minority group. [47 CFR 76.504]

minority group
Includes Blacks, Hispanics, American Indians, Alaska Natives, Asians, and Pacific Islanders. [47 USC 309]

minutes-of-use
A unit of measurement expressed as either holding time or conversation time. [47 CFR Pt. 36, App.]

minutes-of-use-kilometers
The product of (a) the number of minutes-of-use and (b) the average route kilometers of circuits involved. [47 CFR Pt. 36, App.]

miscellaneous common carriers
Communications common carriers which are not engaged in the business of providing either a public landline message telephone service or public message telegraph service. [47 CFR 21.2]

mission-affecting outage
Defined as an outage that is deemed critical to national security/emergency preparedness (NS/EP) operations of the affected facility by the National Communications System member agency operating the affected facility. [47 CFR 63.100]

MMDS
Multichannel multipoint distribution service [47 CFR 74.832]

mobile earth station [1]
An earth station in the mobile-satellite service intended to be used while in motion or during halts at unspecified points. (RR) [47 CFR 2.1]

mobile earth station [2]
An earth station intended to be used while in motion or during halts at unspecified points. [47 CFR 21.2, 25.201]

mobile relay station
A base station in the mobile service authorized to retransmit automatically on a mobile service frequency communications which originate on the transmitting frequency of the mobile station. [47 CFR 90.7]

mobile repeater station
A mobile station authorized to retransmit automatically on a mobile service frequency, communications to or from hand-carried transmitters. [47 CFR 90.7]

mobile-satellite service
A radiocommunication service (1) between mobile earth stations and one or more space stations, or between space stations used by this service; or (2) between mobile earth stations by means of one or more space stations. Note: This service may also include feeder links necessary for its operation. (RR) [47 CFR 2.1, 25.201]

mobile service [1]

A radio communication service carried on between mobile stations or receivers and land stations, and by mobile stations communicating among themselves, and includes (A) both one-way and two-way radio communication services, (B) a mobile service which provides a regularly interacting group of base, mobile, portable, and associated control and relay stations (whether licensed on an individual, cooperative, or multiple basis) for private one-way or two-way land mobile radio communications by eligible users over designated areas of operation, and (C) any service for which a license is required in a personal communications service established pursuant to the proceeding entitled "Amendment to the Commission's Rules to Establish New Personal Communications Services" (GEN docket No. 90-314; ET Docket No. 92-100), or any successor proceeding. [47 USC 153, 47 CFR 20.3]

mobile service [2]

A radiocommunication service between mobile and land stations, or between mobile stations. (CONV) [47 CFR 2.1, 24.5, 26.4, 87.5]

mobile service [3]

A service of radiocommunication between mobile and land stations, or between mobile stations. [47 CFR 5.4, 21.2, 90.7]

mobile station [1]

A radio communication station capable of being moved and which ordinarily does move. [47 USC 153, 47 CFR 21.2]

mobile station [2]

A station in the mobile service intended to be used while in motion or during halts at unspecified points. (RR) [47 CFR 2.1, 5.4, 24.5, 26.4, 90.7]

mobile station [3]

One or more transmitters that are capable of operation while in motion. [47 CFR 22.99]

mobile station [4]

One or more units which transmit while moving or during temporary stops at unspecified points. [47 CFR 95.23]

model I facility

A station operating in the 1605-1705 kHz band featuring full-time operation with stereo, competitive technical quality, 10 kW daytime power, 1 kW nighttime power, non-directional antenna (or a simple directional antenna system), and separated by 400-800 km from other co-channel stations. [47 CFR 73.14]

model II facility

A station operating in the 535-1605 kHz band featuring full-time operation, competitive technical quality, wide area daytime coverage with nighttime coverage at least 15% of the daytime coverage. [47 CFR 73.14]

monochrome transmission

The transmission of television signals which can be reproduced in gradations of a single color only. [47 CFR 73.681]

motion picture producer

Motion picture producer refers to a person or organization engaged in the production of filming of motion pictures. [47 CFR 74.801]

MSAs

Metropolitan statistical areas [47 CFR 22.909]

MTAs

Major trading areas [47 CFR 24.102]

MTA-based license or MTA license

A license authorizing the right to use a specified block of SMR spectrum within one of the 51 Major Trading Areas ("MTAs"), as embodied in Rand NcNally's Trading Area System MTA Diskette and geographically represented in the map contained in Rand McNally's Commercial Atlas & Marketing Guide (the "MTA Map.") The MTA Listings, the MTA Map and the Rand McNally/AMTA license agreement are available for public inspection at the Wireless Telecommunications Bureau's public reference room, Room 628, 1919 M Street NW., Washington, DC 20554. [47 CFR 90.7]

MTF

Maintenance test facility [47 CFR 87.5]

MTS
> Message telecommunications service [47 CFR 61.42]

multi-center exchange
> An exchange area in which are located two or more local central office buildings or wire centers. [47 CFR Pt. 36, App.]

multichannel multipoint distribution service
> Those multipoint distribution service channels that use the frequency band 2596 MHZ to 2644 MHZ and associated response channels. [47 CFR 21.2]

multichannel television sound (MTS)
> Any system of aural transmission that utilizes aural baseband operation between 15 kHz and 120 kHz to convey information or that encodes digital information in the video portion of the television signal that is intended to be decoded as audio information. [47 CFR 73.681]

multichannel video programming distributor [1]
> A person such as, but not limited to, a cable operator, a multichannel multi point distribution service, a direct broadcast satellite service, or a television receive-only satellite program distributor, who makes available for purchase, by subscribers or customers, multiple channels of video programming. [47 USC 522]

multichannel video programming distributor [2]
> An entity engaged in the business of making available for purchase, by subscribers or customers, multiple channels of video programming. Such entities include, but are not limited to, a cable operator, a multichannel multipoint distribution service, a direct broadcast satellite service, a television receive-only satellite program distributor, and satellite master antenna television system operator, as well as buying groups or agents of all such entities. [47 CFR 76.1000, 76.1300]

multilateration LMS system
> A system that is designed to locate vehicles or other objects by measuring the difference of time of arrival, or difference in phase, of signals transmitted from a unit to a number of fixed points or from a number of fixed points to the unit to be located. [47 CFR 90.7]

multi-port equipment
Equipment that has more than one telephone connection with provisions internal to the equipment for establishing transmission paths among two or more telephone connections. [47 CFR 68.3]

multi-satellite link
A radio link between a transmitting earth station and a receiving earth station through two or more satellites, without any intermediate earth station. Note: A multisatellite link comprises one up-link, one or more satellite-to-satellite links and one down-link. (RR) [47 CFR 2.1]

multiple address system (MAS)
A multiple address radio system is a point-to-multipoint communications system, either one-way or two-way, utilizing frequencies listed in §94.65(a)(1) and serving a minimum of four remote stations. If a master station is part of the multiple address system, the remote stations must be scattered over the service area in such a way that two or more point-to-point systems would be needed to serve those remotes. [47 CFR 94.3]

multiple operation
Broadcasting by a station on one frequency over two or more transmitters simultaneously. If a station uses the same frequency simultaneously on each of two (three, etc.) transmitters for an hour, it uses one frequency-hour and two (three, etc.) transmitter-hours. [47 CFR 73.701]

multiplex transmission
The simultaneous transmission of two or more signals within a single channel. Multiplex transmission as applied to FM broadcast stations means the transmission of facsimile or other signals in addition to the regular broadcast signals. [47 CFR 73.310]

multiplex transmission (aural)
A subchannel added to the regular aural carrier of a television broadcast station by means of frequency modulated subcarriers. [47 CFR 73.681]

multiplier
Ownership interests that are held indirectly by any party through one or more intervening corporations will be determined by successive

multiplication of the ownership percentages for each link in the vertical ownership chain and application of the relevant attribution benchmark to the resulting product, except that if the ownership percentage for an interest in any line in the chain exceeds 50 percent or represents actual control, it shall be treated as if it were a 100 percent interest. [47 CFR 90.814]

multipoint distribution service

     A one-way domestic public radio service rendered on microwave frequencies from a fixed station transmitting (usually in an omnidirectional pattern) to multiple receiving facilities located at fixed points. [47 CFR 21.2]

multipoint distribution service response station

     A fixed station operated at an MDS receive location to provide communications with the associated station in the Multipoint Distribution Service. [47 CFR 21.2]

multiunit installations

     (1) In multiunit premises existing as of August, 13, 1990, the demarcation point shall be determined in accordance with the local carrier's reasonable and nondiscriminatory standard operating practices. Provided, however, that where there are multiple demarcation points within the multiunit premises, a demarcation point for a customer shall not be further inside the customer's premises than a point twelve inches from where the wiring enters the customer's premises.

     (2) In multiunit premises in which wiring is installed after August 13, 1990, including additions, modifications and rearrangements of wiring existing prior to that date, the telephone company may establish a reasonable and nondiscriminatory practice of placing the demarcation point at the minimum point of entry. If the telephone company does not elect to establish a practice of placing the demarcation point at the minimum point of entry, the multiunit premises owner shall determine the location of the demarcation point or points. The multiunit premises owner shall determine whether there shall be a single demarcation point location for all customers or separate such locations for each customer. Provided, however, that where there are multiple demarcation points within the multiunit premises, a demarcation point for a customer shall not be further inside the customer's premises than a point 30 cm (12 in) from where the wiring enters the customer's premises. [47 CFR 68.3]

N

NARBA

North American Regional Broadcasting Agreement [47 CFR 73.21]

narrowband PCS

PCS services operating in the 901-902 MHZ, 930-931 MHZ, and 940-941 MHZ bands. [47 CFR 24.5]

narrowband PCS service areas

Nationwide, regional, major trading areas (MTAs) and basic trading areas (BTAs) as defined below. MTAs and BTAs are based on the Rand McNally 1992 Commercial Atlas & Marketing Guide, 123rd Edition, at pages 38-39 ("BTA/MTA Map"). Rand McNally organizes the 50 states and the District of Columbia into 47 MTAs and 487 BTAs. The BTA/MTA Map is available for public inspection at the Office of Engineering and Technology's Technical Information Center, Room 7317, 2025 M Street, NW, Washington, DC.

(a) The nationwide service area consists of the fifty states, the District of Columbia, American Samoa, Guam, Northern Mariana Islands, Puerto Rico, and United States Virgin Islands.

(b) The regional service areas are defined as follows:

(1) Region 1 (Northeast): The Northeast Region consists of the following MTAs: Boston-Providence, Buffalo-Rochester, NewYork, Philadelphia, and Pittsburgh.

(2) Region 2 (South): The South Region consists of the following MTAs: Atlanta, Charlotte-Greensboro-Greenville-Raleigh, Jacksonville, Knoxville, Louisville-Lexington-Evansville, Nashville, Miami-Fort Lauderdale, Richmond-Norfolk, Tampa-St. Petersburg-Orlando, and Washington-Baltimore; and, Puerto Rico and United States Virgin Islands.

(3) Region 3 (Midwest): The Midwest Region consists of the following MTAs: Chicago, Cincinnati-Dayton, Cleveland, Columbus, Des

Moines-Quad Cities, Detroit, Indianapolis, Milwaukee, Minneapolis-St. Paul, and Omaha.

(4) Region 4 (Central): The Central Region consists of the following MTAs: Birmingham, Dallas-Fort Worth, Denver, El Paso-Albuquerque, Houston, Kansas City, Little Rock, Memphis-Jackson, New Orleans-Baton Rouge, Oklahoma City, San Antonio, St. Louis, Tulsa, and Wichita.

(5) Region 5 (West): The West Region consists of the following MTAs: Honolulu, Los Angeles-San Diego, Phoenix, Portland, Salt Lake City, San Francisco-Oakland-San Jose, Seattle (including Alaska), and Spokane-Billings; and, American Samoa, Guam, and the Northern Mariana Islands.

(c) The MTA service areas are based on the Rand McNally 1992 *Commercial Atlas & Marketing Guide*, 123rd Edition, at pages 38-39, with the following exceptions and additions:

(1) Alaska is separated from the Seattle MTA and is licensed separately.

(2) Guam and the Northern Mariana Islands are licensed as a single MTA-like area.

(3) Puerto Rico and the United States Virgin Islands are licensed as a single MTA-like area.

(4) American Samoa is licensed as a single MTA-like area.

(d) The BTA service areas are based on the Rand McNally 1992 *Commercial Atlas & Marketing Guide*, 123rd Edition, at pages 38-39, with the following additions licensed separately as BTA-like areas: American Samoa; Guam; Northern Mariana Islands; Mayagüez/Aguadilla-Ponce, Puerto Rico; San Juan, Puerto Rico; and the United States Virgin Islands. The Mayagüez/Aguadilla-Ponce BTA-like service area consists of the following municipios: Adjuntas, Aguada, Aguadilla, Añasco, Arroyo, Cabo Rojo, Coamo, Guánico, Guayama, Guayanilla, Hormigueros, Isabela, Jayuya, Juana Díaz, Lajas, Las Marías, Maricao, Maunabo, Mayagüez, Moca, Patillas, Peñuelas, Ponce, Quebradillas, Rincón, Sabana Grande, Salinas, San Germán, Santa Isabel, Villalba, and Yauco. The San Juan BTA-like service area consists of all other municipios in Puerto Rico. [47 CFR 24.102]

national audience reach

The total number of television households in the Arbitron Area of Dominant Influence (ADI) markets in which the relevant stations are located divided by the total nation television households as measured by ADI data at the time of a grant, transfer or assignment of a license. For purposes of making this calculation, UHF television stations shall be attributed with 50 percent of the television households in their ADI market. Where the relevant application forms require a showing with respect to audience reach and the application relates to an area where Arbitron ADI market data are unavailable, then the applicant shall make a showing as to the number of television households in its market. Upon such a showing, the Commission shall make a determination as to the appropriate audience reach to be attributed to the applicant. [47 CFR 73.3555]

National Communications System (NCS) [1]
That organization established by the President in Executive Order No. 12472, "Assignment of National Security and Emergency Preparedness Telecommunications Functions," (April 3, 1984, 49 FR 13471 (1984). [47 CFR 64.1604]

National Communications System (NCS) [2]
A confederation of federal departments, agencies and entities established by Presidential Memorandum of August 21, 1963 and reaffirmed by Executive Order No. 12472, "Assignment of National Security and Emergency Preparedness Telecommunications Functions." April 3, 1984. [47 CFR Pt. 216, App.]

National Coordinating Center (NCC)
The joint telecommunications industry—federal government operation established by the NCS to assist in the initiation, coordination, restoration and reconstitution of NSEP telecommunication services or facilities. [47 CFR Pt. 216, App.]

national educational programming supplier
Any qualified noncommercial educational television station, other public telecommunications entities, and public or private educational institutions. [47 USC 335]

National Geodetic Reference System (NGRS)
The name given to all geodetic control data contained in the National Geodetic Survey (NGS) data base. (Source: National Geodetic Survey, U.S. Department of Commerce) [47 CFR 26.4]

national radio quiet zone

The area in Maryland, Virginia and West Virginia bounded by 39° 15'N on the north, 78° 30'W on the east, 37° 30'N on the south and 80° 30'W on the west. [47 CFR 97.3]

National Security Emergency Preparedness (NSEP)

Telecommunication services which are used to maintain a state of readiness or to respond to and manage any event or crisis (local, national, or international), which causes or could cause injury or harm to the population, damage to or loss of property, or degrades or threatens the NSEP posture of the United States. These services fall into two specific categories, Emergency NSEP and Essential NSEP, and are assigned priority levels pursuant to section 9 of this appendix. [47 CFR 64.1604, 47 CFR Pt. 216, App.]

National Security Emergency Preparedness (NSEP) Treatment

The provisioning of a telecommunications service before others based on the provisioning priority level assigned by the Manager, NCS, in accordance with this directive. [47 CFR Pt. 216, App.]

navigable waters

As used in reference to waters of the United States, its territories and possessions, the waters shoreward of the baseline of its territorial sea and internal waters as contained in 33 CFR 2.05-25. [47 CFR 80.5, 90.7]

navigational communications

Safety communications pertaining to the maneuvering of vessels or the directing of vessel movements. Such communications are primarily for the exchange of information between ship stations and secondarily between ship stations and coast stations. [47 CFR 80.5]

NBC

National Broadcasting Company [47 CFR 11.43]

NBDP

Narrow-band direct printing [47 CFR 80.121]

NBS

National Bureau of Standards [47 CFR 73.1207]

NCE

Noncommercial educational [47 CFR 76.55]

NCE-FM

Noncommercial educational FM [47 CFR 73.506]

NCS

National communications system [47 CFR 63.100]

NECA

National Exchange Carrier Association [47 CFR 36.611]

necessary bandwidth [1]

The calculated spectral width of an emission. Calculations are made using procedures set forth in part 2 of this chapter. The bandwidth so calculated is considered to be the minimum necessary to convey information at the desired rate with the desired accuracy. [47 CFR 22.99]

necessary bandwidth [2]

For a given class of emission, the minimum value of the occupied bandwidth sufficient to ensure the transmission of information at the rate and with the quality required for the system employed, under specified conditions. Emissions useful for the good functioning of the receiving equipment, as for example, the emission corresponding to the carrier of reduced carrier systems, shall be included in the necessary bandwidth. [47 CFR 74.401, 94.3]

necessary bandwidth of emission

For a given class of emission, the width of the frequency band that is just sufficient to ensure the transmission of information at the rate and with the quality required under specified conditions. (RR) [47 CFR 2.1]. Note: The necessary bandwidth for an emission may be calculated using the formulas in §2.202 of this chapter. [47 CFR 21.2]

negative transmission

Where a decrease in initial light intensity causes an increase in the transmitted power. [47 CFR 73.681]

net investment

Allowable original cost investment in Accounts 2001 through 2003, 1220 and 1402 that has been apportioned to interstate and foreign services pursuant to the *Separations Manual* from which depreciation, amortization and other reserves attributable to such investment that has been apportioned to interstate and foreign services pursuant to the *Separations Manual* have been subtracted and to which working capital that is attributable to interstate and foreign services has been added. [47 CFR 69.2]

net weekly circulation

The number of noncable television households that viewed the station for 5 minutes or more during the entire week, expressed as a percentage of the total noncable television households in the survey area. [47 CFR 76.5]

network

For purposes of the must carry rules, a commercial television network is an entity that offers programming on a regular basis for 15 or more hours per week to at least 25 affiliates in 10 or more states. [47 CFR 76.55]

network element

A facility or equipment used in the provision of a telecommunications service. Such term also includes features, functions, and capabilities that are provided by means of such facility or equipment, including subscriber numbers, databases, signaling systems, and information sufficient for billing and collection or used in the transmission, routing, or other provision of a telecommunications service. [47 USC 153]

network port

An equipment port of registered protective circuitry which port faces the telephone network. [47 CFR 68.3]

network program

Any program delivered simultaneously to more than one broadcast station regional or national, commercial or noncommercial. [47 CFR 76.5]

new service offering

A tariff filing that provides for a class or sub-class of service not previously offered by the carrier involved and that enlarges the range of service options available to ratepayers. [47 CFR 61.3]

news

Information that is about current events or that would be of current interest to the public. [47 CFR 0.466]

NEXRAD

Government Next Generation Weather Radar [47 CFR 2.106]

NG

Non-government [47 CFR 2.106]

NGRS

National Geodetic Reference System [47 CFR 24.5]

NGS

National Geodetic System [47 24.5]

NIC

National Information Center [47 CFR 11.16]

nighttime

The period of time between local sunset and local sunrise. [47 CFR 73.14]

900 MHZ SMR MTA-based license or MTA license

A license authorizing the right to use a specified block of 900 MHZ SMR spectrum within one of the 47 Major Trading Areas ("MTAs"), as embodied in Rand McNally's Trading Areas System MTA Diskette and geographically represented in the map contained in Rand McNally's Commercial Atlas & Marketing Guide (the "MTA Map"), with the following exceptions and additions:

(1) Alaska is separated from the Seattle MTA and is licensed separately.

(2) Guam and the Northern Mariana Islands are licensed as a single MTA-like area.

(3) Puerto Rico and the United States Virgin Islands are licensed as a single MTA-like area.

(4) American Samoa is licensed as a single MTA-like area.

The MTA map is available for public inspection in the Office of Engineering and Technology's Technical Information Center, Room 7317, 2025 M Street NW., Washington, DC. [47 CFR 90.7]

NN

Non-participating national [47 CFR 11.18]

nodal station

The central or controlling station in a radio system operating on point-to-multipoint frequencies in the 2.5, 10.6, and 18 GHz bands. [47 CFR 94.3]

nominal power

The antenna input power less any power loss through a dissipative network and, for directional antennas, without consideration of adjustments specified in paragraphs (b)(1) and (b)(2) of 73.51 of the rules. However, for AM broadcast applications granted or filed before June 3, 1985, nominal power is specified in a system of classifications which include the following values: 50 kW, 25 kW, 10 kW, 5 kW, 2.5 kW, 1 kW, 0.5 kW, and 0.25 kW. The specified nominal power for any station in this group of stations will be retained until action is taken on or after June 3, 1985, which involves a change in the technical facilities of the station. [47 CFR 73.14]

nonattributable equity

(i) For corporations, voting stock or non-voting stock that includes no more than twenty-five percent of the total voting equity, including the right to vote such stock through a voting trust or other arrangement;

(ii) For partnerships, joint ventures and other non-corporate entities, limited partnership interests and similar interests that do not afford the power to exercise control of the entity. [47 CFR 24.720]

nonbroadcast telecommunications facilities

Includes, but is not limited to, cable television systems, communications satellite systems and related terminal equipment, and other modes of transmitting, emitting, or receiving images and sounds or intelligence by means of wire, radio, optical, electromagnetic, or other means. [47 USC 395]

noncommercial communications

Communication between coast stations and ship stations other than commercial transport ships, or between ship stations aboard other than commercial transport ships which pertain to the needs of the ship. [47 CFR 80.5]

noncommercial educational broadcast station (public broadcast station)

A television or radio broadcast station which (A) under the rules and regulations of the Commission in effect on the effective date of this paragraph, is eligible to be licensed by the Commission as a noncommercial educational radio or television broadcast station and which is owned and operated by a public agency or nonprofit private foundation, corporation, or association; or (B) is owned and operated by a municipality and which transmits only noncommercial programs for educational purposes. [47 USC 397]

noncommercial FM translator

An FM broadcast translator station which rebroadcasts the signals of a noncommercial educational FM radio broadcast station. [47 CFR 74.1201]

noncommercial telecommunications entity

Any enterprise which (A) is owned and operated by a state, a political or special purpose subdivision of a State, a public agency, or a nonprofit private foundation, corporation, or association; and (B) has been organized primarily for the purpose of disseminating audio or video noncommercial educational and cultural programs to the public by means other than a primary television or radio broadcast station, including, but not limited to, coaxial cable, optical fiber, broadcast translators, cassettes, discs, microwave, or laser transmission thought the atmosphere. [47 USC 397]

noncommercial scientific institution

An institution that is not operated on a commercial basis as that term is referenced in paragraph (a) (4) of this section, and which is operated solely for the purpose of conducting scientific research the results of which are not intended to promote any particular product or industry. [47 CFR 0.466]

noncoordinatable PCS device

A PCS device that is capable of randomly roaming and operating in geographic areas containing incumbent microwave facilities such that operation of the PCS device will potentially cause harmful interference to the incumbent microwave facilities. [47 CFR 15.303]

nondominant carrier

A carrier not found to be dominant. [47 CFR 61.3]

nonfederal financial support

The total value of cash and the fair market value of property and services (including, to the extent provided in the second sentence of this paragraph, the personal services of volunteers), received--

(A) as gifts, grants, bequests, donations, or other contributions for the construction or operation of noncommercial educational broadcast stations, or for the production, acquisition, distribution, or dissemination of educational television or radio programs, and related activities, from any source other than (i) the United States or any agency or instrumentality of the United States; or (ii) any public broadcasting entity; or

(B) as gifts, grants, donations, contributions, or payments from any state, or any educational institution, for the construction or operation of noncommercial educational broadcast stations or for the production, acquisition, distribution, or dissemination of educational television or radio programs, or payments in exchange for services or materials with respect to the provision of educational or instructional television or radio programs.

Such term includes the fair market value of personal services or volunteers, and computed using the valuation standards established by the Corporation and approved by the Comptroller General pursuant to Section 396(g)(5), but only with respect to such services provided to public telecommunications entities after such standards are approved by the Comptroller General and only, with respect to such an entity in a fiscal year, to the extent that the value of the services does not exceed 5 percent of the total non-federal financial support of the entity in such fiscal year. [47 USC 397]

nonmultilateration LMS system

A system that employs any of a number of non-multilateration technologies to transmit information to and/or from vehicular units. [47 CFR 90.7]

nonprofit

As applied to any foundation, corporation, or association, a foundation, corporation, or association, no part of the net earnings of which inures, or may lawfully inure, to the benefit of any private shareholder or individual. [47 USC 397]

nonselectable transponder

A transponder whose coded response is displayed on any conventional radar operating in the appropriate band. [47 CFR 80.5]

nonsystem premises wiring

Wiring which is used with one and two-line business and residence services, located at the subscriber's premises. [47 CFR 68.3]

nonvoice, nongeostationary mobile-satellite service

A mobile-satellite service reserved for use by non-geostationary satellites in the provision of non-voice communications which may include satellite links between land earth stations at fixed locations. [47 CFR 2.1, 25.201]

normal business hours

Those hours during which most similar businesses in the community are open to serve customers. In all cases, "normal business hours" must include some evening hours at least one night per week and/or some weekend hours. [47 CFR 76.309]

normal operating conditions

Those service conditions which are within the control of the cable operator. Those conditions which are *not* within the control of the cable operator include, but are not limited to, natural disasters, civil disturbances, power outages, telephone network outages, and severe or unusual weather conditions. Those conditions which *are* ordinarily within the control of the cable operator include, but are not limited to, special promotions, pay-perview events, rate increases, regular peak or seasonal demand periods, and maintenance or upgrade of the cable system. [47 CFR 76.309]

November season

That portion of any year commencing at 0100 g.m.t. on the first Sunday in November and ending at 0100 g.m.t. on the first Sunday in March. [47 CFR 73.701]

NP

National primary [47 CFR 11.18]

NPR

National Public Radio [47 CFR 11.43]

NPT

National periodic test [47 CFR 11.31]

NPTs

New product tiers [47 CFR 76.987]

NRAO

National Radio Astronomy Observatory [47 CFR 22.369]

NRRO

Naval Radio Research Observatory [47 CFR 22.369]

NRSC

National Radio Systems Committee [47 CFR 11.51]

NRZ

Non-return-to-zero [47 CFR 73.682]

NSEP

National Security and Emergency Preparedness [47 CFR 0.183]

NSEP treatment

The provisioning of a telecommunication service before others based on the provisioning of a telecommunication service before others based on the provisioning priority level assigned by the Executive Office of the President. [47 CFR 64.1604]

NTIA

National Telecommunications and Information Administration [47 CFR 1.1114]

NTIS

National Technical Information Service [47 CFR 0.434]

nuclear ship

A ship provided with a nuclear power plant. [47 USC 153]

number portability

The ability of users of telecommunications services to retain, at the same location, existing telecommunications numbers without impairment of quality, reliability, or convenience when switching from one telecommunications carrier to another. [47 USC 153]

NWS

National Weather Service [47 CFR 11.20]

O

occupied bandwidth [1]

The width of a frequency band such that, below the lower and above the upper frequency limits, the mean powers emitted are each equal to a specified percentage Beta/2 of the total mean power of a given emission. Note: Unless otherwise specified by the CCIR for the appropriate class of emission, the value of Beta/2 should be taken as 0.5%. (RR) [47 CFR 2.1]

occupied bandwidth [2]

The measured spectral width of an emission. The measurement determines occupied bandwidth as the difference between upper and lower frequencies where 0.5% of the emission power is above the upper frequency and 0.5% of the emission power is below the lower frequency. [47 CFR 22.99]

occupied bandwidth [3]

The frequency bandwidth such that, below its lower and above its upper frequency limits, the mean powers radiated are each equal to 0.5 percent of the total mean power radiated by a given emission [47 CFR 23.1, 74.401]. In some cases, for example, multichannel frequency-division systems, the percentage 0.5 percent may lead to certain difficulties in the practical application of the definitions of occupied and necessary bandwidth; in such cases a different percentage may prove useful. [47 CFR 94.3]

OET

Office of Engineering and Technology [47 CFR 0.241]

off-premises line simulator circuit

A load impedance for connection, in lieu of an off-premises station line, to PBX (or similar) telephone system loop start circuits (Figure 68.3(f) during testing. The schematic diagram of Figure 68.3(f) is illustrative of the type of circuit which will be required; alternative implementations may be used provided that the same dc voltage and current characteristics and ac impedance characteristics will be presented to the equipment under test as are presented in the illustrative schematic diagram. When used, the simulator shall be operated over the entire range of loop resistances as indicated in Figure 68.3(f), and with the indicated polarities. Whenever loop current is changed, sufficient time shall be allocated for the current to reach a steady-state condition before continuing testing. [47 CFR 68.3]

off-premises station interface

The point of connection between PBX telephone systems (or similar systems) and telephone company private line communication facilities used to access registered station equipment located off the premises. Equipment leads at this interface are limited to telephone tip and ring leads (designated T(OPS) and R(OPS)) where the PBX employs loop-start signaling at the interface. Unless otherwise noted, all T(OPS) and R(OPS) leads shall be treated as telephone connections for purposes of fulfilling registration conditions. [47 CFR 68.3]

Office of Personnel Management Regulations

The regulations (5 CFR, part 735) on employee responsibilities and conduct issued by the Office of Personnel Management on October 1, 1965, in implementation of Executive Order 11222. [47 CFR 19.735-102]

officer

Includes furloughed, pensioned, and superannuated officers. [47 CFR 41.1]

officer (director)

Shall include the duties, or any of the duties, ordinarily performed by a director, president, vice president, secretary, treasurer, or other officer of a carrier, such as general counsel, general solicitor, general attorney, comptroller, general auditor, general manager, general commercial manager, chief engineer, general superintendent, general land and tax agent, or chief purchasing agent. [47 CFR 62.2]

official responsibility

As used in this section is defined in 5 CFR 737.7(b) [47 CFR 1.25]

offshore central transmitter

A fixed transmitter in the Offshore Radiotelephone Service that provides service to offshore subscriber stations. [47 CFR 22.99]

offshore radiotelephone service

A radio service in which common carriers are authorized to offer and provide radio telecommunication services for hire to subscribers on structures in the offshore coastal waters of the Gulf of Mexico. [47 CFR 22.99]

offshore subscriber station

One or more fixed and/or mobile transmitters in the Offshore Radiotelephone Service that receive service from offshore central transmitters. [47 CFR 22.99]

OMB

Office of Management and Budget [47 CFR 0.408]

on-board communication station

A low-powered mobile station in the maritime mobile service intended for use for internal communications on board a ship, or between a ship and its lifeboats and life-rafts during lifeboat drills or operations, or for communication within a group of vessels being towed or pushed, as well as for line handling and mooring instructions. (RR) [47 CFR 2.1, 80.5]

on-board repeater

A radio station that receives and automatically retransmits signals between on-board communication stations. [47 CFR 80.5]

1.544 Mbps digital CO 4-wire interface

A 4-wire digital interface between digital terminal equipment and a digital transmission facility operating at 1.544 Mbps connecting to a serving central office. [47 CFR 68.3]

1.544 Mbps digital service

A full-time dedicated private line circuit used for the transmission of digital signals at a speed of 1.544 Mbps. [47 CFR 68.3]

1.6/2.4 GHz mobile-satellite service

A mobile-satellite service that operates in the 1610-1626.5 MHZ and 2483.5-2500 MHZ frequency bands, or in any portion thereof. [47 CFR 25.201]

one-port equipment

Equipment which has either exactly one telephone connection, or a multiplicity of telephone connections arranged so that no transmission among such telephone connections, within the equipment, is intended. [47 CFR 68.3]

open sea

The water area of the open coast seaward of the ordinary low-water mark, or seaward of inland waters. [47 CFR 80.5]

operating taxes

Operating taxes include all taxes in Account 7200. [47 CFR 69.2]

operational communications

Communications concerning the technical and programming operation of a broadcast station and its auxiliaries. [47 CFR 74.401]

operational fixed station [1]

A fixed station, not open to public correspondence, operated by entities that provide their own radiocommunication facilities in the private land mobile, maritime or aviation services. [47 CFR 80.5]

operational fixed station [2]

A fixed station, not open to public correspondence, operated by and for the sole use of persons operating their own radiocommunication facilities in the public safety, industrial, land transportation, marine, or aviation services. [47 CFR 87.5, 90.7]

operational-fixed station [3]

A fixed station not open to public correspondence, operated by and for the sole use of those persons or agencies operating their own radio communication facilities. This term includes all stations licensed in the fixed service under this part. [47 CFR 94.3]

operations

The term denoting the general classifications of services rendered to the public for which separate tariffs are filed, namely exchange, state toll and interstate toll. [47 CFR Pt. 36, App.]

operator

(A) For the purpose of Parts II and III of Title III of this Act, a person on a ship of the United States holding a radio operator's license of the proper class as prescribed and issued by the Commission.

(B) For the purpose of Part II of Title III of this Act, a person on a foreign ship holding a certificate as such of the proper class complying with the provisions of the radio regulations annexed to the International Telecommunication Convention in force, or complying with an agreement or treaty between the United States and the country in which the ship is registered. [47 USC 153]

operator services

Any interstate telecommunications service initiated from an aggregator location that includes, as a component, any automatic or live assistance to a consumer to arrange for billing or completion, or both, of an interstate telephone call through a method other than (A) automatic completion with billing to the telephone from which the call originated; or (B) completion through an access code used by the consumer, with billing to an account previously established with the carrier by consumer. [47 USC 226, 47 CFR 64.708]

operator system

A stored program electronic system associated with one or more toll switching systems which provides centralized traffic service position functions for several local offices at one location. [47 CFR Pt. 36, App.]

operator trunks

A general term, ordinarily applied to trunks between manually operated switchboard positions and local dial central offices in the same wire center. [47 CFR Pt. 36, App.]

OPS

Off-premises station [47 CFR 68.3]

optimum working frequency (FOT)

The highest frequency which is returned by ionospheric radio propagation to the surface of the earth for a particular path and time of day for 90 percent of the days of the reference month. [47 CFR 73.701]

OR

Off-route [47 CFR 87.5]

orbit

The path relative to a specified frame of reference, described by the centre of mass of a satellite or other object in space subjected primarily to natural forces, mainly the force of gravity. (RR) [47 CFR 2.1]

order

That portion of its action wherein the Commission announces its judgment. This should be distinguished from the "memorandum opinion" or other material which often accompany and explain the order. [47 CFR 1.106, 1.115]

order, decision, report, or action

Does not include an initial, tentative, or recommended decision to which exceptions may be filed as provided in Section 409(b).[47 USC 155]

original cost or cost

As applied to telecommunications plant, rights of way and other intangible property, means the actual money cost of (or the current money value of any consideration other than money exchanged for) property at the time when it was first dedicated to use by a regulated telecommunications entity, whether the accounting company or by predecessors.

For the application of this definition to property acquired from predecessors, see §32.2000(b)(1) of Subpart C. Note also the definition of Cost in this section. [47 CFR 32.9000]

origination cablecasting

Programming (exclusive of broadcast signals) carried on a cable television system over one or more channels and subject to the exclusive control of the cable operator. [47 CFR 76.5]

origination of service

(A) Origination of a service that is switched in a Class 4 switch or an interexchange switch that performs an equivalent function ends when the transmission enters such switch and termination of such a service begins when the transmission leaves such a switch, except that;

(1) Switching in a Class 4 switch or transmission between Class 4 switches that is not deemed to be interexchange for purposes of the Modified Final Judgement entered August 24, 1982, in *United States v Western Electric Co.*, D.C. civil Action No. 82-0192, will be "origination" or "termination" for purposes of this part; and (2) *Origination* and *termination* does not include the use of any part of a line, trunk or switch that is not owned or leased by a telephone company.

(B) Origination of any service other than a service that is switched in a Class 4 switch or a switch that performs an equivalent function ends and "termination" of any such service begins at a point of demarcation that corresponds with the point of demarcation that is used for a service that is switched in a Class 4 switch or a switch that performs an equivalent function. [47 CFR 69.2]

other area

The area where the coverage contour of an FM translator station extends beyond the protected contour of the primary station (*i.e.*, predicted 0.5 mV/m contour for commercial Class B stations, predicted 0.7 mV/m contour for commercial Class B1 stations, and predicted 1 mV/m contour for all other classes of stations). [47 CFR 74.1201]

other eligible system

A system comprised of microwave radio channels in the Multipoint Distribution Service and Multichannel Multipoint Distribution Services (as defined in §21.2 of this chapter, and, on a part-time basis, in the Instructional Television Fixed Service (as defined in §74.901 of this chapter) that delivers multichannel television service over the air to subscribers. [47 CFR 78.5]

other participating carrier

A carrier subject to the Act that publishes a tariff containing rates and regulations applicable to the portion or through service it furnishes in conjunction with another subject carrier. [47 CFR 61.3]

other programming service

Information that a cable operator makes available to all subscribers generally. [47 USC 522]

other service

Any wire or radio communications service provided using any of the facilities of a cable operator that are used in the provision of cable service. [47 USC 551]

out-of-band emission

Emission on a frequency or frequencies immediately outside the necessary bandwidth which results from the modulation process, but excluding spurious emissions. (RR) [47 CFR 2.1]

outage

A significant degradation in the ability of a customer to establish and maintain a channel of communications as a result of failure or degradation in the performance of a carrier's network. [47 CFR 63.100]

outage which potentially affects a major airport

An outage that disrupts 50% or more of the air traffic control links or other FAA communications links to any major airport, any outage that has caused an ARTCC or major airport to lose its radar, any ARTCC or major airport outage that is likely to be of media interest, any outage that causes a loss of both primary and backup facilities at any ARTCC or major airport, and any outage to an ARTCC or major airport that is deemed important by the FAA as indicated by FAA inquiry to the carrier management personnel. [47 CFR 63.100]

outage which potentially affects a 911 special facility

An outage which disrupts more than 25% of the lines to any PSAP without providing automatic rerouting to an alternative PSAP. [47 CFR 63.100]

output power

The radio frequency output power of a transmitter's final radio frequency stage as measured at the output terminal while connected to a load of the impedance recommended by the manufacturer. [47 CFR 90.7]

own

To have a direct or indirect equity interest (or the equivalent thereof) of more than 10 percent of an entity, or the right to more than 10 percent of the gross revenues of an entity under a revenue sharing or royalty agreement. [47 USC 274]

P

pager
A small radio receiver designed to be carried by a person and to give an aural, visual or tactile indication when activated by the reception of a radio signal containing its specific code. It may also reproduce sounds and/or display messages that were also transmitted. Some pagers also transmit a radio signal acknowledging that a message has been received. [47 CFR 22.99]

paging
A one-way communications service from a base station to mobile or fixed receivers that provide signaling or information transfer by such means as tone, tone-voice, tactile, optical readout, etc. [47 CFR 90.7]

paging and radiotelephone service
A radio service in which common carriers are authorized to offer and provide paging and radiotelephone service for hire to the general public. This service was formerly titled Public Land Mobile Service. [47 CFR 22.99]

paging receiver
A unit capable of receiving the radio signals from a base station for the bearer to hear a page (someone's name or other identifier said in order to find, summon or notify him/her) spoken by the base station operator. [47 CFR 95.25]

paging service
Transmission of coded radio signals for the purpose of activating specific pagers; such transmissions may include messages and/or sounds. [47 CFR 22.99]

PARS
Paging and radiotelephone service [47 CFR 22.602]

partial network station

A commercial television broadcast station that generally carries in prime time more than 10 hours of programming per week offered by the three major national television networks, but less than the amount specified in paragraph (j) of this section. [47 CFR 76.5]

participated personally and substantially

As used in this section is defined in 5 CFR 737.5(d). [47 CFR 1.25]

particular commission matter involving a specific party

When used in paragraphs (c) and (d) of this section, any judicial or other proceeding, application, request for a ruling or other determination, contract, claim controversy, investigation, charge accusation, arrest or other particular matter involving a specific party or parties in which the Commission is a party or has a direct and substantial interest. [47 CFR 1.25]

partitioned cellular market

A cellular market with two or more authorized cellular systems on the same channel block during the five year build-out period, as a result of settlements during initial licensing or contract(s) between the licensee of the first cellular system and the licensee(s) of the subsequent systems. See §22.947(b) [47 CFR 22.99]

partitioned service area (PSA)

The area within the coterminous boundaries of one or more counties or other geopolitical subdivisions, drawn from a BTA, to which an authorization holder may provide Multipoint Distribution Service or the area remaining in a BTA upon partitioning any portion of that BTA. This area excludes the protected service areas of incumbent MDS stations and the registered receive sites of previously authorized and proposed ITFS stations. [47 CFR 21.2]

partitioned service area authorization holder

The individual or entity authorized by the Commission to provide Multipoint Distribution Service to the population of a partitioned service area. [47 CFR 21.2]

party

Any person who participates in a proceeding by the timely filing of a petition for rule making, comments on a notice of proposed rule making, a petition for reconsideration, or responsive pleadings in the manner prescribed by this subpart. The term does not include those who submit letters, telegrams or other informal materials. [47 CFR 1.400]

passenger

Any person carried on board a ship or vessel except (1) the officers and crew actually employed to man and operate the ship, (2) persons employed to carry on the business of the ship, and (3) persons on board a ship when they are carried, either because of the obligation laid upon the master to carry shipwrecked, distressed or other persons in like or similar situations or by reason of any circumstance over which neither the master, the owner nor the charterer (if any) has control. [47 USC 153]

passenger ship

A ship that carries or is licensed or certificated to carry more than twelve passengers. [47 USC 153]

passenger ship safety certificate

A certificate issued by the Commandant of the Coast Guard after inspection of a passenger ship which complies with the requirements of the Safety Convention. [47 CFR 80.5]

passive equity

(1) For corporations, non-voting stock or stock that includes no more than fifteen percent of the voting equity; (2) for partnerships, joint ventures and other non-corporate entities, limited partnership interests and similar interests that do not afford the power to exercise control of the entity. [47 CFR 24.230]

passive repeater

A passive antenna element or elements, located to reflect radiation from or redirect radiation to a directional transmitting and/or receiving antenna in a horizontal or near horizontal plane to a horizontal or near horizontal plane. [47 CFR 94.3]

passive sensor
A measuring instrument in the earth exploration-satellite service or in the space research service by means of which information is obtained by reception of radio waves of natural origin. (RR) [47 CFR 2.1]

paying agency
The agency employing the individual and authorizing the payment of his or her current pay. [47 CFR 1.1901]

pay-per-call [1]
Any service:
(1) In which any person provides or purports to provide (i) audio information or audio entertainment produced or packaged by such person; (ii) access to simultaneous voice conversation services; or (iii) any service, including the provision of a product, the charges for which are assessed on the basis of the completion of the call;
(2) For which the caller pays a per call or per-time-interval charge that is greater than, or in addition to, the charge for transmission of the call; and
(3) Which is accessed through use of a 900 number;
(4) Provided, however, such term does not include directory services provided by a common carrier or its affiliate or by a local exchange carrier or its affiliate, or any service the charge for which is tariffed, or any service for which users are assessed charges only after entering into a presubscription or comparable arrangement with the provider of such service. [47 CFR 64.1501]

pay-per-call services [2]
Any service--
(A) in which any person provides or purports to provide (i) audio information or audio entertainment produced or packaged by such person; (ii) access to simultaneous voice conversation services; or (iii) any service, including the provision of a product, the charges for which are assessed on the basis of the completion of the call;
(B) for which the caller pays a per-call or per-time-interval charge that is greater than, or in addition to, the charge for transmission of the call; and
(C) which is accessed through use of a 900 telephone number or other prefix or area code designated by the Commission in accordance with subsection (b)(5). [47 USC 228]

pay-per-call services [3]

Has the meaning provided in section 228(i) of the Communications Act of 1934, except that the Commission by rule may, notwithstanding subparagraphs (B) and (C) of section 228(i)(1) of such Act, extend such definition to other similar services providing audio information or audio entertainment if the Commission determines that such services are susceptible to the unfair and deceptive practices that are prohibited by the rules prescribed pursuant to section 201(a). [47 USC 701]

payphone service

The provision of public or semi-public pay telephones, the provision of inmate telephone service in correctional institutions, and ancillary services. [47 USC 276]

PBX

Private branch exchange [47 CFR 68.3]

PCI

Price cap index [47 CFR 61.3]

PCS

Personal communications services [47 CFR 15.301]

peak envelope power [1]

TP averaged during 1 RF cycle at the highest crest of the modulation envelope. [47 CFR 95.669]

peak envelope power (PEP) [2]

The average power supplied to the antenna transmission line by a transmitter during one RF cycle at the crest of the modulation envelope taken under normal operating conditions. [47 CFR 97.3]

peak envelope power (of a radio transmitter) [3]

The average power supplied to the antenna transmission line by a transmitter during one radio frequency cycle at the crest of the modulation envelope taken under normal operating conditions. (RR) [47 CFR 2.1, 87.5]

peak power

The power over a radio frequency cycle corresponding in amplitude to synchronizing power. [47 CFR 73.681]

peak power of a radio transmitter

The mean power supplied to the antenna during one radio frequency cycle at the highest crest of the modulation envelope, taken under conditions of normal operation. [47 CFR 5.4]

peak transmit power

The peak power output as measured over an interval of time equal to the frame rate or transmission burst of the device under all conditions of modulation. Usually this parameter is measured as a conducted emission by direct connection of a calibrated test instrument to the equipment under test. If the device cannot be connected directly, alternative techniques acceptable to Commission may be used. [47 CFR 15.303]

PEP

Primary entry point system [47 CFR 11.31]

PEP

Peak envelope power [47 CFR 2.106]

percentage modulation [1]

The ratio of the actual frequency deviation to the frequency deviation defined as 100% modulation, expressed in percentage. For FM broadcast stations, a frequency deviation of ±75 kHz is defined as 100% modulation. [47 CFR 73.310]

percentage modulation [2]

As applied to frequency modulation, the ratio of the actual frequency deviation to the frequency deviation defined as 100% modulation expressed in percentage. For the aural transmitter of TV broadcast stations, a frequency deviation of ± 25 kHz is defined as 100% modulation. [47 CFR 73.681]

percentage modulation (amplitude)

In a positive direction:

$$M = \frac{MAX\text{-}C}{c} \times 100$$

In a negative direction:

$$M = \frac{C\text{-}MIN}{c} \times 100$$

Where:

M = Modulation level in percent.

MAX = Instantaneous maximum level of the modulated radio frequency envelope.

MIN = Instantaneous minimum level of the modulated radio frequency envelope.

C = (Carrier) level of radio frequency envelope without modulation. [47 CFR 73.14]

perimeter protection system

A field disturbance sensor that employs RF transmission lines as the radiating source. These RF transmission lines are installed in such a manner that allows the system to detect movement within the protected area. [47 CFR 15.3]

period (of a satellite)

The time elapsing between two consecutive passages of a satellite through a characteristic point on its orbit. (RR) [47 CFR 2.1]

peripheral device

An input/output unit of a system that feeds data into and/or receives data from the central processing unit of a digital device. Peripherals to a digital device include any device that is connected external to the digital device, any device internal to the digital device that connects the digital device to an external device by wire or cable, and any circuit board or card designed for interchangeable mounting, internally or externally, that increases the operating or processing speed of a digital device, *e.g.*, "turbo cards" and "enhancement boards". Examples of peripheral devices include terminals, printers, external floppy disk drives and other data storage devices, video monitors, keyboards, control cards, interface boards, external memory expansion cards and other input/output devices that may or may not contain digital circuitry. However, an internal device that contains the central processing unit of a digital device is not a peripheral even though such a device may connect to an external keyboard or other components. [47 CFR 15.3]

periscope antenna system
An antenna system which involves the use of a passive reflector to deflect radiation from or to a directional transmitting or receiving antenna which is oriented vertically. [47 CFR 21.2, 94.3]

permanent
Whenever a station is to transmit from a single location, the station location is *permanent* and the location must be shown on the application. [47 CFR 80.39]

permanently affixed
That the required nameplate data is etched, engraved, stamped, indelibly printed or otherwise permanently marked. Alternatively, the required information may be permanently marked on a nameplate of metal, plastic, or other material fastened to the enclosure by welding, riveting, etc., or with a permanent adhesive. Such a nameplate must be able to last the expected lifetime of the equipment in the environment in which the equipment will be operated and must not be readily detachable. [47 CFR 68.300]

permissible interference
Observed or predicted interference which complies with quantitative interference and sharing criteria contained in these [international Radio] Regulations or in CCIR Recommendations or in special agreements as provided for in these Regulations. (RR) [47 CFR 2.1]

PERP
Peak effective radiated power [47 CFR 2.1511]

person [1]
An individual, partnership, association, joint-stock company, trust, or corporation. [47 USC 153, 47 CFR 5.4, 90.7, 94.3]

person [2]
An individual, partnership, association, joint stock company, trust, corporation, or state or local governmental entity. [47 USC 394]

person [3]

An individual, a corporation, a company, an association, a firm, a partnership, a society, a joint stock company, or any other organization or institution. [47 CFR 19.735-102]

person [4]

An individual, partnership, committee, association, corporation, labor organization, and any other organization or group of persons. [47 CFR 64.803]

person in the television industry

A television network, any entity which produces programming (including theatrical motion pictures) for telecasting or telecasts programming, the National Cable Television Association, the Association of Independent Television Stations, Incorporated, the National Association of Broadcasters, the Motion Picture Association of America, the Community Antenna Television Association, and each of the networks' affiliate organizations, and shall include any individual acting on behalf of such person. [47 USC 303c]

personal communications services (PCS) devices [unlicensed]

Intentional radiators operating in the frequency bands 1910-1930 MHZ and 2390-2400 MHZ that provide a wide array of mobile and ancillary fixed communication services to individuals and businesses. [47 CFR 15.303]

personal computer

An electronic computer that is marketed for use in the home, notwithstanding business applications. Such computers are considered Class B digital devices. Computers which use a standard TV receiver as a display device or meet all of the following conditions are considered examples of personal computers: (1) marketed through a retail outlet or direct mail order catalog; (2) notices of sale or advertisements are distributed or directed to the general public or hobbyist users rather than restricted to commercial users; (3) operates on a battery or 120 volt electrical supply.

If the responsible party can demonstrate that because of price or performance the computer is not suitable for residential or hobbyist use, it may request that the computer be considered to fall outside the scope of this definition for personal computers. [47 CFR 15.3]

personal communications services (PCS) devices [unlicensed]
Intentional radiators operating in the frequency bands 1910-1930 MHZ and 2390-2400 MHZ that provide a wide array of mobile and ancillary fixed communication services to individuals and businesses. [47 CFR 15.303]

personal net worth
The market value of all assets (real and personal, tangible and intangible) owned by an individual, less all liabilities (including personal guarantees) owed by the individual in his individual capacity or as a joint obligor. [47 CFR 24.320]

personal wireless services
Commercial mobile services, unlicensed wireless services, and common carrier wireless exchange access services. [47 USC 332]

personal wireless service facilities
Facilities for the provision of personal wireless services. [47 USC 332]

personally identifiable information
Does not include any record of aggregate data which does not identify particular persons. [47 USC 551]

physician
For the purpose of this part, a person who is licensed to practice in a place where the amateur service is regulated by the FCC, as either a Doctor of Medicine (M.D.) or a Doctor of Osteopathy (D.O.) [47 CFR 97.3]

PIC
Primary interexchange carrier [47 CFR 64.1100]

pilot
A federal pilot required by 46 U.S.C. 764, a state pilot required under the authority of 46 U.S.C. 211, or a registered pilot required by 46 U.S.C. 216. [47 CFR 80.5]

pilot subcarrier [1]
A subcarrier that serves as a control signal for use in the reception of FM stereophonic sound broadcasts. [47 CFR 73.310]

pilot subcarrier [2]

A subcarrier used in the reception of TV stereophonic aural or other subchannel broadcasts. [47 CFR 73.681]

plant retired

Plant which has been removed, sold, abandoned, destroyed, or otherwise withdrawn from service. [47 CFR 32.9000]

plate modulation

The modulation produced by introduction of the modulating wave into the plate circuit of any tube in which the carrier frequency wave is present. [47 CFR 73.14]

PLC

Power line carrier [47 CFR 2.106]

pleading

Any written notice, motion, petition, request, opposition, reply, brief, proposed findings, exceptions, memorandum of law, or other paper filed with the Commission in a hearing proceeding. It does not include exhibits or documents offered in evidence. §See 1.356. [47 CFR 1.204]

PN

Participating national [47 CFR 11.18]

point of communication

A specific location designated in the license to which a station is authorized to communicate for the transmission of public correspondence. [47 CFR 23.1]

point-to-point ITFS station

An ITFS station that transmits a highly directional signal from a fixed transmitter location to a fixed receive location. [47 CFR 74.901]

point-to-point microwave radio service

A domestic public radio service rendered on microwave frequencies by fixed stations between points which lie within the United States or between points to its possessions or to points in Canada or Mexico. [47 CFR 21.2]

point-to-point telephone station

A fixed station authorized for radiotelephone communication. [47 CFR 23.1]

polarization

The direction of the electric field as radiated from the transmitting antenna. [47 CFR 73.681]

pole attachment

Any attachment by a cable television system to a pole, duct, conduit, or right-of-way owned or controlled by a utility. [47 CFR 1.1402]

port (see harbor)

port operations communications

Communications in or near a port, in locks or in waterways between coast stations and ship stations or between ship stations, which relate to the operational handling, movement and safety of ships and in emergency to the safety of persons. [47 CFR 80.5]

port operations service

A maritime mobile service in or near a port, between coast stations and ship stations, or between ship stations, in which messages are restricted to those relating to the operational handling, the movement and the safety of ships and, in emergency, to the safety of persons. (RR) Note: Messages which are of a public correspondence nature shall be excluded from this service. (RR) [47 CFR 2.1]

port station

A coast station in the port operations service. (RR) [47 CFR 2.1]

portable ship station

A ship station which includes a single transmitter intended for use upon two or more ships. [47 CFR 80.5]

power

Whenever the power of a radio transmitter, etc. is referred to it shall be expressed in one of the following forms, according to the class of emission, using the arbitrary symbols indicated: (1) peak envelope power (PX or pX);

(2) mean power (PY or pY); (3) carrier power (PZ or pZ). Note 1: For different classes of emission, the relationships between peak envelope power, mean power and carrier power, under the conditions of normal operation and of no modulation, are contained in CCIR Recommendations which may be used as a guide. Note 2: For use in formulae, the symbol "p" denotes power expressed in watts and the symbol "P" denotes power expressed in decibels relative to a reference level. (RR) [47 CFR 2.1]

power connections
        The connection between commercial power and any transformer, power supply rectifier, converter or other circuitry associated with registered terminal equipment or registered protective circuitry. The following are not power connections : (A) connections between registered terminal equipment or registered protective circuitry and sources of nonhazardous voltages (see §68.306(b)(4) for a definition of non-hazardous voltages); (B) conductors which distribute any per within registered terminal equipment or within registered protective circuitry; (C) green wire ground (the grounded conductor of a commercial power circuit which is UL-identified by a continuous green color). [47 CFR 68.3]

power line carrier systems
        An unintentional radiator employed as a carrier current system used by an electric power utility entity on transmission lines for protective relaying, telemetry, etc. for general supervision of the power system. The system operates by the transmission of radio frequency energy by conduction over the electric power transmission lines of the system. The system does not include those electric lines which connect the distribution substation to the customer or house wiring. [47 CFR 15.3]

PPC
        Proof-of-passing certificate [47 CFR 13.3]

PRA
        Paperwork Reduction Act [47 CFR 0.408]

preexisting entity; existing investor
        A preexisting entity is an entity that was operating and earning revenues for at least two years prior to December 31, 1994. An existing investor is a person or entity that was an owner of record of a preexisting

entity's equity as of November 10, 1994, and any person or entity acquiring *de minimus* equity holdings in a preexisting entity after that date. [47 CFR 24.720]

premises

A dwelling unit, other building or a legal unit of real property such as a lot on which a dwelling unit is located, as determined by the telephone company's reasonable and nondiscriminatory standard operating practices. [47 CFR 68.3]

premium channel

Any pay service offered on a per channel or per program basis, which offers movies rated by the Motion Picture Association of America as X, NC-17, or R. [47 USC 544]

preoperational expenses

All nonconstruction costs incurred by new telecommunications entities before the date on which they begin providing service to the public, and all nonconstruction costs associated with expansion of existing entities before the date on which such expanded capacity is activated, except that such expenses shall not include any portion of the salaries of any personnel employed by an operating public telecommunications entity. [47 USC 397]

presentation

Any communication directed to the merits or outcome of a proceeding. Excluded from this term is a communication which is inadvertently or casually made, or a communication which is an inquiry or request for information relating solely to the status of a proceeding. A status inquiry which states or implies a preference for a particular party or position in a proceeding, or which states why timing is important to a particular party, or which in any other manner is intended as a means, direct or indirect, to address the merits or outcome, or influence the timing, or proceeding is a presentation. [47 CFR 1.1202]

presubscribed provider of operator services

The interstate provider of operator services to which the consumer is connected when the consumer places a call using a provider of operator services without dialing an access code. [47 USC 226, 47 CFR 64.708]

presubscription (comparable arrangement)
A contractual agreement in which:

(1) The service provider clearly and conspicuously discloses to the consumer all material terms and conditions associated with the use of the service, including the service provider's name and address, a business telephone number which the consumer may use to obtain additional information or to register a complaint, and the rates for the service;

(2) The service provider agrees to notify the consumer of any future rate changes;

(3) The consumer agrees to use the service on the terms and conditions disclosed by the service provider;

(4) The service provider requires the use of an identification number or other means to prevent unauthorized access to the service by nonsubscribers; and

(5) Provided, however, that disclosure of a credit or charge card number, along with authorization to bill that number, made during the course of a call to an information service shall constitute a presubscription or comparable arrangement if the credit or charge card is subject to the dispute resolution procedures of the Truth in Lending Act and Fair Credit Billing Act, as amended, 15 U.S.C. 1601 *et seq.* No other action taken by a consumer during the course of a call to an information service, for which charges are assessed, can create a presubscription or comparable arrangement. [47 CFR 64.1501]

price cap index (PCI)
An index of costs applying to carriers subject to price cap regulation, which index is calculated for each basket pursuant to §61.44 or 61.45. [47 CFR 61.3]

price cap regulation
A method of regulation of dominant carriers provided in §§61.41 through 61.49. [47 CFR 61.3, 69.2]

price cap tariff
Any tariff filing involving a service that is within a price cap basket, or that requires calculations pursuant to §61.44, 61.45, 61.46, or 61.47. [47 CFR 61.3]

primary radar

A radiodetermination system based on the comparison of reference signals with radio signals reflected from the position to be determined. (RR) [47 CFR 2.1]

primary service area

The service area of a broadcast station in which the groundwave is not subject to objectionable interference or objectionable fading. [47 CFR 73.14]

primary station [1]

The television broadcast station which provides the programs and signals being retransmitted by a television broadcast translator station. [47 CFR 74.701]

primary station [2]

The FM radio broadcast station radiating the signals which are retransmitted by an FM broadcast translator station or an FM broadcast booster station. [47 CFR 74.1201]

prime time [1]

One that has network exhibition during the hours of 7-11 pm eastern and pacific time or 6-10 p.m. central and mountain time. [47 CFR 73.662]

prime time [2]

The 5-hour period from 6 to 11 p.m., local time, except that in the central time zone the relevant period shall be between the hours of 5 and 10 p.m., and in the mountain time zone each station shall elect whether the period shall be 6 to 11 p.m. or 5 to 10 p.m. [47 CFR 76.5]

principal headend

(1) The headend, in the case of a cable system with a single headend or, (2) in the case of a cable system with more than one headend, the principal headend designated by the cable operator, except that such designation shall not undermine or evade the requirements of subpart D of this part. The designation of a principal headend shall be made by May 3, 1993, and each cable system shall place in its public file the location of its designated principal headend by June 17, 1993, as provided in §76.302. Except for good cause, an operator may not change its choice of principal headend. [47 CFR 76.5]

priority action [1]

Assignment, revision, revocation, or revalidation by the Executive Office of the President of a priority level associated with an NSEP telecommunications service. [47 CFR 64.1604]

priority action [2]

The assignment, revision, revocation, or revalidation by the Manager, NCS, in accordance with this directive, of a priority level associated with an NSEP telecommunications service. [47 CFR Pt. 21 6, App.]

priority level

The level that may be assigned to an NSEP telecommunications service specifying the order in which provisioning or restoration of the service is to occur relative to other NSEP and/or non-NSEP telecommunication services. Authorized priority levels are designated (highest to lowest) "E, "1," "2," "3," "4,"and "5" for provisioning and "1," "2," "3," "4," and "5" for restoration. [47 CFR Pt. 216, App., 47 CFR 64. 1604]

priority level assignment

The priority level(s) designated for the provisioning and/or restoration of a particular NSEP telecommunications service. [47 CFR Pt. 216, App., 47 CFR 64.1604]

privacy indicator

Information, contained in the calling party number parameter of the call set-up message associated with an interstate call on an Signaling System 7 network, that indicates whether the calling party authorizes presentation of the calling party number to the called party. [47 CFR 64.1600]

private aircraft station

A mobile station on board an aircraft not operated as an air carrier. A station on board an air carrier aircraft weighing less than 12,500 pounds maximum certified take-off gross weight may be licensed as a private aircraft station. [47 CFR 87.5]

private carrier

An entity licensed in the private services and authorized to provide communications service to other private services on a commercial basis. [47 CFR 90.7, 94.3]

private coast station

A coast station, not open to public correspondence, which serves the operational, maritime control and business needs of ships. [47 CFR 80.5]

private financial gain

Shall not include the gain resulting to any individual for the private use in such individual's dwelling unit of any programming for which the individual has not obtained authorization for that use. [47 USC 605]

private line

A line that is used exclusively for an interexchange service other than MTS, WATS or an MTS-WATS equivalent service, including a line that is used at the closed end of an FX WATS or CCSA service or any service that is substantially equivalent to a CCSA service. [47 CFR 69.2]

private line channel

Telephone company dedicated facilities and channel equipment used in furnishing private line service from the telephone network for the exclusive use of a particular party or parties. [47 CFR 68.3]

private line service [1]

A service whereby facilities for communication between two or more designated points are set aside for the exclusive use or availability for use of a particular customer and authorized users during stated periods of time. [47 CFR 21.2]

private line service [2]

A service for communications between specified locations for a continuous period or for regularly recurring periods at stated hours. [47 CFR Pt. 36, App.]

private mobile radio service

A mobile service that is neither a commercial mobile radio service nor the functional equivalent of a service that meets the definition of commercial mobile radio service. Private mobile radio service includes the following:

(a) Not-for-profit land mobile radio and paging services that serve the licensee's internal communications needs as defined in part 90 of this chapter. Shared-use, cost-sharing, or cooperative arrangements, multiple licensed systems that use third party managers or users combining resources to meet compatible needs for specialized internal communications facilities in compliance with the safeguards of §90.179 of this chapter are presumptively private mobile radio services;

(b) Mobile radio service offered to restricted classes of eligible users. This includes the following services: Public Safety Radio Services; Special Emergency Radio Service; Industrial Radio Services (excluding Business Radio Services that offer customers for-profit interconnected services); Land Transportation Radio Services; and Radiolocation Services;

(c) 220-222 MHZ land mobile service and Automatic Vehicle Monitoring systems (part 90 of this chapter) that do not offer interconnected service or that are not-for-profit; and

(d) Personal Radio Services under part 95 of this chapter (General Mobile Services, Radio Control Services, and Citizens Band Radio Services); Maritime Service Stations (excluding Public Coast stations) (part 80 of this chapter); and Aviation Service Stations (part 87 of this chapter). [47 CFR 20.3]

private mobile service

Any mobile service (as defined in Section 3) that is not a commercial mobile service or the functional equivalent of a commercial mobile service, as specified by regulation by the Commission. [47 USC 332]

private NSEP telecommunication services

Those non-common carrier telecommunication services including private line, virtual private line, and private switched network services. [47 CFR Pt. 216, App., 47 CFR 1604]

private radio services

Private land mobile radio services and other communications services characterized by the Commission in its rules as private radio services. [47 USC 605, 47 CFR 68.3]

private viewing

The viewing for private use in an individual's dwelling unit by means of equipment, owned or operated by such individual, capable of receiving satellite cable programming directly from a satellite. [47 USC 605]

productivity factor
>An adjustment factor used to make annual adjustments to the Price Cap Index to reflect the margin by which a carrier subject to price cap regulation is expected to improve its productivity relative to the economy as a whole. [47 CFR 61.3]

program
>Includes any complete program or part thereof. [47 CFR 74.184]

program origination
>For purposes of this part, program origination shall be any transmissions other than the simultaneous retransmission of the programs and signals of a TV broadcast station. Origination shall include locally generated television program signals and program signals obtained via video recordings (tapes and discs), microwave, common carrier circuits, or other sources. [47 CFR 74.701]

program related data signal
>A signal, consisting of a series of pulses representing data, which is transmitted simultaneously with and directly related to the accompanying television program. [47 CFR 73.681]

program solely produced by a network
>A program in which the network, directly or through a production entity it owns or controls, is the sole copyright owner, has full financial responsibility and full business and production control. [47 CFR 73.662]

proof of performance measurements or antenna proof of performance measurements
>The measurements of field strengths made to determine the radiation pattern or characteristics of an AM directional antenna system. [47 CFR 73.14]

protected system premises wiring requiring acceptance testing for imbalance
>Premises wiring which is electrically behind registered (or grandfathered) equipment, system components or circuitry which assure that electrical contact between the wiring and commercial power wiring will not result in hazardous voltages at the telephone network interface. [47 CFR 68.3]

protection ratio

The minimum value of the wanted-to-unwanted signal ratio, usually expressed in decibels, at the receiver input determined under specified conditions such that a specified reception quality of the wanted signal is achieved at the receiver output. (RR) [47 CFR 2.1]

provider of direct broadcast satellite service

(i) A licensee for a Ku-band satellite system under Part 100 of Title 47 of the Code of Federal Regulations; or (ii) any distributor who controls a minimum number of channels (as specified by Commission regulation) using a Ku-banc fixed service satellite system for the provision of video programming directly to the home and licensed under Part 25 of Title 47 of the Code of Federal Regulations. [47 USC 335]

provider of operator services

Any common carrier that provides operator services or any other person determined by the Commission to be providing operator services. [47 USC 226, 47 CFR 64.708]

provisioning

The act of supplying telecommunications service to a user, including all associated transmission, wiring, and equipment. As used herein, "provisioning" and "initiation" are synonymous and include altering the state of an existing priority service or capability. [47 CFR Pt. 216, App., 47 CFR 1604]

PSA

Partitioned service area [47 CFR 21.2]

pseudorandom sequence

A sequence of binary data which has some of the characteristics of a random sequence but also has some characteristics which are not random. It resembles a true random sequence in that the one bits and zero bits of the sequence are distributed randomly throughout every length, N, of the sequence and the total numbers of the one and zero bits in that length are approximately equal. It is not a true random sequence, however, because it consists of a fixed number (or length) of coded bits which repeats itself exactly whenever that length is exceeded, and because it is generated by a fixed algorithm from some fixed initial state. [47 CFR 2.1]

PSN

Public switched network [47 CFR 90.353]

PSRA

Presunrise service authorization [47 CFR 73.99]

PSSA

Presunset [47 CFR 73.1670]

PSSA

Postsunset service authorization [47 CFR 73.99]

public broadcasting entity

The Corporation, any licensee or permittee of a public broadcast station, or any nonprofit institution engaged primarily in the production, acquisition, distribution, or dissemination of educational and cultural television or radio programs. [47 USC 397]

public coast station

A coast station that offers radio communication common carrier services to ship radio stations. [47 CFR 80.5]

public correspondence [1]

Any telecommunication which the offices and stations must, by reason of their being at the disposal of the public, accept for transmission. (CONV) [47 CFR 2.1, 80.5]

public correspondence [2]

Any telecommunications which the offices and stations, by reason of their being at the disposal of the public, must accept for transmission. [47 CFR 5.4, 21.2]

public, educational, or governmental access facilities

(A) Channel capacity designated for public, educational, or governmental use; and (B) facilities and equipment for the use of such channel capacity. [47 USC 522]

public harm

Public harm must begin immediately and cause direct and actual damage to property or to the health or safety of the general public, or diversion of law enforcement or other public health and safety authorities from their duties. The public harm will be deemed foreseeable if the licensee could expect with a significant degree of certainty that public harm would occur. [47 CFR 73.1225]

public institutions telecommunications user

An elementary or secondary school, a library, or a health care provider as those terms are defined in this paragraph. [47 USC 254]

public library

Has the same meaning given such term in section 3 of the Library Services and Construction Act. [47 USC 708]

public message service

A service whereby facilities are offered to the public for communication between all points served by a carrier or by interconnected carriers on a non-exclusive message-by-message basis, contemplating a separate connection for each occasion of use. [47 CFR 21.2]

public mobile services [1]

Air-to-ground radiotelephone services, cellular radio telecommunications services, offshore radio, rural radio service, public land mobile telephone service, and other common carrier radio communication services covered by Part 22 of Title 47 of the Code of Federal Regulations. [47 USC 605, 47 CFR 68.3]

public mobile services [2]

Radio services in which common carriers are authorized to offer and provide mobile and related fixed radio telecommunication services for hire to the public. [47 CFR 22.99]

public notice

The date of any of the following events:

(1) For documents in notice and comment rule making proceedings, including summaries thereof, the date of publication in the FEDERAL REGISTER.

(2) For non-rulemaking documents released by the Commission or staff, whether or not published in the FEDERAL REGISTER, the release date. A document is "released" by making the full text available to the press and public in the Commission's Office of Public Affairs. The release date appears on the face of the document.

(3) For rule makings of particular applicability, if the rule making document is to be published in the FEDERAL REGISTER and the Commission so stated in its decision, the date of public notice will commence on the day of the FEDERAL REGISTER publication date. If the decision fails to specify FEDERAL REGISTER publication, the date of public notice will commence on the release date, even if the document is subsequently published in the FEDERAL REGISTER. See Declaratory Ruling, 51 FR 23059 (June 25, 1986).

(4) If the full text of an action document is not to be released by the commission, but a descriptive document entitled "Public Notice" describing the action is released, the date on which the descriptive "Public Notice" describing the action is released, the date on which the descriptive "Public Notice" is released. [47 CFR 1.4]

public switched network

Any common carrier switched network, whether by wire or radio, including local exchange carriers, interexchange carriers, and mobile service providers, that use the North American Numbering Plan in connection with the provision of switched services. [47 CFR 20.3]

public switched NSEP telecommunication services

Those NSEP telecommunication services utilizing public switched networks. Such services may include both interexchange and intraexchange network facilities (*e.g.,* switching systems, interoffice trunks and subscriber loops). [47 CFR Pt. 216, App., 47 CFR 1604]

public telecommunications entity

Any enterprise which (A) is a public broadcast station or a noncommercial telecommunications entity; and (B) disseminates public telecommunications services to the public. [47 USC 397]

public telecommunications facilities

Apparatus necessary for production, interconnection, captioning, broadcast, or other distribution of programming, including, but not limited to, studio equipment, cameras, microphones, audio and video storage or reproduction equipment, or both, signal processors and switchers, towers, antennas, transmitters, translators, microwave equipment, mobile equipment, satellite communications equipment, instructional television fixed service equipment, subsidiary communications authorization transmitting and receiving equipment, cable television equipment, video and audio cassettes and discs, optical fiber communications equipment, and other means of transmitting, emitting, storing, and receiving images and sounds, or intelligence, except that such term does not include the buildings to house such apparatus (other than small equipment shelters which are part of satellite earth stations, translators, microwave interconnection facilities, and similar facilities). [47 USC 397]

public telecommunications network interconnectivity

The ability of two or more public telecommunications networks used to provide telecommunications service to communicate and exchange information without degeneration, and to interact in concert with one another. [47 USC 256]

public telecommunications services

Noncommercial educational and cultural radio and televisiograms, and related noncommercial instructional or informational material that may be transmitted by means of electronic communications. [47 USC 397]

public telephone

A telephone provided by a telephone company through which an end user may originate interstate or foreign telecommunications for which he pays with coins or by credit card, collect or third number billing procedures. [47 CFR 69.2]

public toll station

A public telephone station, located in a community, through which a carrier provides service to the public, and which is connected directly to a toll line operated by such carrier. [47 CFR 63.60]

publication (see tariff publication)

publicly traded corporation with widely dispersed voting power

A business entity organized under the laws of the United States:

(1) Whose shares, debt, or other ownership interests are traded on an organized securities exchange within the United States;

(2) In which no person (i) owns more than 15 percent of the equity; or (ii) possesses, directly or indirectly, through the ownership of voting securities, by contract or otherwise, the power to control the election of more than 15 percent of the members of the board of directors or other governing body of such publicly traded corporation; and

(3) Over which no person other than the management and members of the board of directors or other governing body of such publicly traded corporation, in their capacities as such, has *de facto* control.

(4) The term *person* shall be defined as in section 13(d) of the Securities and Exchange Act of 1934, as amended (15 U.S.C. 78(m)), and shall also include investors that are commonly controlled under the indicia of control set forth in the definition of *affiliate* in paragraphs (1)(2) through (1) of this section. [47 CFR 24.720]

pulsed FM systems

A pulsed FM system is a spread spectrum system in which a RF carrier is modulated with a fixed period and fixed duty cycle sequence. At the beginning of each transmitted pulse, the carrier frequency is frequency modulated causing an additional spreading of the carrier. The pattern of the frequency modulation will depend upon the spreading function which is chosen. In some systems the spreading function is a linear FM chirp sweep, sweeping either up or down in frequency. [47 CFR 2.1]

Q

qualified educational programming source

A programming source which devotes substantially all of its programming to educational or instructional programming that promotes public understanding of mathematics, the sciences, the humanities, and the arts and has a documented annual expenditure on programming exceeding $15,000,000. The annual expenditure on programming means all annual costs incurred by the programming source to produce or acquire programs which are scheduled to be televised, and specifically excludes marketing, promotion, satellite transmission and operational costs, and general administrative costs. [47 USC 532]

qualified individual with handicaps

(1) With respect to any Commission program or activity under which an individual is required to perform services or to achieve a level of accomplishment, an individual with handicaps who meets the essential eligibility requirements and who can achieve the purpose of the program or activity without modifications in the program or activity that the Commission can demonstrate would result in a fundamental alteration in its nature; and

(2) With respect to any other program or activity, an individual with handicaps who meets the essential eligibility requirements for participation in, or receipt of benefits from, that program or activity; and

(3) Qualified handicapped person as that term is defined for purposes of employment in 29 CFR 1613.702(f), which is made applicable to this part by §1.1840. [47 CFR 1.1803]

qualified local noncommercial educational (NCE) television station

A qualified local NCE television station is a qualified NCE television station (1) that is licensed to community whose reference point, as defined in §76.53 is within 80.45 km (50 miles) of the principal headend, as defined in §76.5(pp), of the cable system; or (2) whose Grade B service contour encompasses the principal headend, as defined in §76.5(pp), of the cable system. [47 CFR 76.55]

qualified low power station [1]

Any television broadcast station conforming to the rules established for Low Power Television Stations contained in Part 74 of Title 47, Code of Federal Regulations, only if--

(A) such station broadcasts for a least the minimum number of hours of operation required by the Commission for television broadcast stations under Part 73 of Title 47, Code of Federal Regulations;

(B) such station meets all obligations and requirements applicable to television broadcast stations under Part 73 of Title 47, Code of Federal Regulations, with respect to the broadcast of nonentertainment programming; programming and rates involving political candidates, election issues, controversial issues of public importance, editorials, and personal attacks; programming for children; and equal employment opportunity; and the Commission determines that the provision of such programming by such station would address local news and informational needs which are not being adequately served by full power television broadcast stations because of the geographic distance of such full power stations from the low power station's community of license;

(C) such station complies with interference regulations consistent with its secondary status pursuant to Part 74 of Title 47, Code of Federal Regulations;

(D) such station is located no more than 35 miles from the cable system's headend, and delivers to the principal headend of the cable system an over-the-air signal of good quality, as determined by the Commission;

(E) the community of license of such station and the franchise area of the cable system are both located outside of the largest 160 Metropolitan Statistical Areas, ranked by population, as determined by the Office of Management and Budget on June 30, 1990, and the population of such community of license on such date did not exceed 35,000; and

(F) there is no full power television broadcast station licensed to any community within the county or other political subdivision (of a state) served by the cable system.

Nothing in this paragraph shall be construed to change the secondary status of any low power station as provided in Part 74 of Title 47, *Code of Federal Regulations*, as in effect on the date of enactment of this section. [47 USC 534]

qualified low power station [2]

A qualified low power station is any television broadcast station conforming to the low power television rules contained in part 74 of this chapter, only if:

(1) Such station broadcasts for at least the minimum number of hours of operation required by the Commission for full power television broadcast stations under part 73 of this chapter;

(2) Such station meets all obligations and requirements applicable to full power television broadcast stations under part 73 of this chapter, with respect to the broadcast of nonentertainment programming; programming and rates involving political candidates, election issues, controversial issues of public importance, editorials, and personal attacks; programming for children; and equal employment opportunity; and the Commission determines that the provision of such programming by such station would address local news and informational needs which are not being adequately served by full power television broadcast stations because of the geographic distance of such full power stations from the lower power station's community of license;

(3) Such station complies with interference regulations consistent with its secondary status pursuant to part 74 of this chapter;

(4) Such station is located no more than 56.32 km (35 miles) from the cable system's principal headend, as defined in 76.5(pp), and delivers to that headend an over-the-air signal of good quality;

(5) The community of license of such station and the franchise area of the cable system are both located outside of the largest 160 Metropolitan Statistical Areas, ranked by population, as determined by the Office of Management and Budget on June 30, 1990, and the population of such community of license on such date did not exceed 35,000; and

(6) There is no full power television broadcast station licensed to any community within the county or other equivalent political subdivision (of a state) served by the cable system. [47 CFR 76.55]

qualified minority programming source

A programming source which devotes substantially all of its programming to coverage of minority viewpoints, or to programming directed at members of minority groups, and which is over 50 percent minority-owned, as the term "minority" is defined in Section 309(i)(3)(C)(ii). [47 USC 532]

qualified noncommercial educational (NCE) television station

A qualified NCE television station is any television broadcast station which

(1)(i) Under the rules and regulations of the Commission in effect on March 29, 1990, is licensed by the Commission as an NCE television broadcast station and which is owned and operated by a public agency, nonprofit foundation, corporation, or association; and (ii) Has as its licensee an entity which is eligible to receive a community service grant, or any successor grant thereto, from the Corporation for Public Broadcasting, or any successor organization thereto, on the basis of the formula set forth in section 366(k)(6)(B) of the Communications Act of 1934, as amended; or

(2) Is owned and operated by municipality and transmits noncommercial programs for educational programs for educational purposes, as defined in §73.621 of this chapter, for at least 50 percent of its broadcast week.

(3) This definition includes (i) the translator of any NCE television station with five watts or higher power serving the franchise area, (ii) A full-service station or translator if such station or translator is licensed to a channel reserved for NCE use pursuant to §73.606 of this chapter, or any successor regulations thereto, and (iii) Such stations and translators operating on channels not so reserved but otherwise qualified as NCE stations. Note: For the purposes of §76.55(a), "serving the franchise area" will be based on the predicted protected contour of the NCE translator. [47 CFR 76.55]

qualifying carrier

A telecommunications carrier that (1) lacks economies of scale or scope, as determined in accordance with regulations prescribed by the Commission pursuant to this section; and (2) offers telephone exchange service, exchange access, and any other service that is included in universal service, to all consumers without preference throughout the service area for which such carrier has been designated as a eligible telecommunications carrier under section 214(e). [47 USC 259]

qualifying investor; qualifying minority and/or woman investor

(1) A qualifying investor is a person who is (or holds an interest in) a member of the applicant's (or licensee's) control group and whose gross revenues and total assets, when aggregated with those of all other attributable investors and affiliates, do not exceed the gross revenues and total assets limits specified in§24.709(a) or §24.715(a), or, in the case of an applicant (or

licensee) that is a small business, do not exceed the gross revenues limit specified in paragraph (b) of this section.

(2) A qualifying minority and/or woman investor is a person who is a qualifying investor under paragraph (n)(1), who is (or holds an interest in) a member of the applicant's (or licensee's) control group and who is a member of a minority group or a woman and a United States citizen. [47 CFR 24.720]

question pool

All current examination questions for a designated written examination element. [47 CFR 13.3, 97.3]

question set

A series of examination questions on a given examination selected from the current question pool. [47 CFR 13.3, 97.3]

quiet zones

Quiet zones are those areas where it is necessary to restrict radiation so as to minimize possible impact on the operations of radio astronomy or other facilities that are highly sensitive to radio frequency interference. [47 CFR 21.113]

R

R
>Route [47 CFR 87.5]

R/C
>Radio control radio service [47 CFR 95.669]

R/C transmitter
>A transmitter that operates or is intended to operate at a station authorized in the R/C.   [47 CFR 95.669]

R(OPS)
>Ring off-premises station [47 CFR 68.3]

RACES (radio amateur civil emergency service)
>A radio service using amateur stations for civil defense communications during periods of local, regional or national civil emergencies. [47 CFR 97.3]

RACON
>Radiobeacon [47 CFR 80.5]

racon station
>A radionavigation land station which employs a racon. A racon (radar beacon) is a transmitter-receiver associated with a fixed navigational mark, which when triggered by a radar, automatically returns a distinctive signal which can appear on the display of the triggering radar, providing range, bearing and identification information. [47 CFR 87.5]

radar

A radiodetermination system based on the comparison of reference signals with radio signals reflected, or retransmitted, from the position to be determined. (RR) [47 CFR 2.1, 87.5]

radar beacon (RACON) [1]

A transmitter-receiver associated with a fixed navigational mark which, when triggered by radar, automatically returns a distinctive signal which can appear on the display of the triggering radar, providing range, bearing and identification information. (RR) [47 CFR 2.1]

radar beacon (RACON) [2]

A receiver transmitter which, when triggered by a radar, automatically returns a distinctive signal which can appear on the display of the triggering radar, providing range, bearing and identification information. [47 CFR 80.5]

radiation

The outward flow of energy from any source in the form of radio waves. (RR) [47 CFR 2.1]

radio

A general term applied to the use of radio waves. (CONV) [47 CFR 2.1]

radio altimeter

Radionavigation equipment, on board an aircraft or spacecraft, used to determine the height of the aircraft or spacecraft above the earth's surface or another surface. [47 CFR 2.1, 87.5]

radio astronomy

Astronomy based on the reception of radio waves of cosmic origin. (RR) [47 CFR 2.1]

radio astronomy service

A service involving the use of radio astronomy. (RR) [47 CFR 2.1]

radio astronomy station

A station in the radio astronomy service. (RR) [47 CFR 2.1]

radiobeacon station

A station in the radionavigation service the emissions of which are intended to enable a mobile station to determine its bearing or direction in relation to radiobeacon station. (RR) [47 CFR 2.1, 87.5]

radio call box

A transmitter used by the public to request fire, police, medical, road service, or other emergency assistance. [47 CFR 90.7]

radio common carrier

A telecommunications common carrier that provides radio communications services but is not engaged in the business of providing landline local exchange telephone service. [47 CFR 22.99]

radio communication (communication by radio)

The transmission by radio of writing, signs, signals, pictures, and sounds of all kinds, including all instrumentalities, facilities, apparatus and services (among other things, the receipt, forwarding, and delivery of communications) incidental to such transmission. [47 USC 153]

radiocommunication [1]

Telecommunication by means of radio waves. (CONV) [47 CFR 2.1]

radiocommunication [2]

Any telecommunication by means of Hertzian waves. [47 CFR 21.2]

radiocommunication service

A service as defined in the Section involving the transmission, emission and/or reception of radio waves for specific telecommunication purposes. Note: In these [International] Radio Regulations, unless otherwise stated, any radiocommunication service relates to terrestrial radiocommunication. (RR) [47 CFR 2.1]

radio control service

Has the meaning given it by the Commission by rule. [47 USC 403]

radio determination [1]

The determination of the position, velocity and/or other characteristics of an object, or the obtaining of information relating to these parameters, by means of the propagation properties of radio waves. (RR) [47 CFR 2.1]

radio determination [2]

The determination of position, or the obtaining of information relating to position, by means of the propagation of radio waves. [47 CFR 90.7]

radiodetermination-satellite service

A radiocommunication service for the purpose of radiodetermination involving the use or one of more space stations. This service may also include feeder links necessary for its own operation. (RR) [47 CFR 2.1, 25.201]

radiodetermination service

A radiocommunication service which uses radiodetermination. Radiodetermination is the determination of the position, velocity and/or other characteristics of an object, or the obtaining of information relating to these parameters, by means of the propagation of radio waves. A station in this service is called a radiodetermination station. [47 CFR 87.5]

radiodetermination station

A station in the radiodetermination service. (RR) [47 CFR 2.1]

radio direction-finding

Radiodetermination using the reception of radio waves for the purpose of determining the direction of a station or object. (RR) [47 CFR 2.1]

radio direction-finding station

A radiodetermination station using radio direction-finding. (RR) [47 CFR 2.1]

radiofacsimile

A system of radiocommunication for the transmission of fixed images, with or without half-tones, with a view a view to their reproduction in a permanent form. [47 CFR 90.7]

radio frequency (RF) energy [1]

Electromagnetic energy at any frequency in the radio spectrum between 9 kHz and 3,000,000 MHZ. [47 CFR 15.3]

radio frequency (RF) energy [2]

Electromagnetic energy at any frequency in the radio spectrum from 9 kHz to 3 THz (3,000 GHz). [47 CFR 18.107]

radiolocation

Radiodetermination used for purposes other than those of radionavigation. (RR) [47 CFR 2.1, 90.7]

radiolocation mobile station

A station in the radiolocation service intended to be used while in motion or during halts at unspecified points. (RR) [47 CFR 2.1]

radiolocation service

A radiodetermination service for the purpose of radiolocation. (RR) [47 CFR 2.1, 87.5]

radionavigation

Radiodetermination used for the purposes of navigation, including obstruction warning. [47 CFR 2.1, 90.7]

radionavigation land station

A station in the radionavigation service not intended to be used while in motion. (RR) [47 CFR 2.1]

radionavigation land test stations

A radionavigation land station which is used to transmit information essential to the testing and calibration of aircraft navigational aids, receiving equipment, and interrogators at predetermined surface locations. The Maintenance Test Facility (MTF) is used primarily to permit maintenance testing by aircraft radio service personnel. The Operational Test Facility (OTF) is use primarily to permit the pilot to check a radionavigation system aboard the aircraft prior to takeoff. [47 CFR 87.5]

radionavigation mobile station

A station in the radionavigation service intended to be used while in motion or during halts at unspecified points. (RR) [47 CFR 2.1]

radionavigation-satellite service

A radiodetermination-satellite service used for the purpose of radio-navigation. This service may also include feeder links necessary for its operation. (RR) [47 CFR 2.1]

radionavigation service

A radiodetermination service for the purpose of radionavigation. (RR) [47 CFR 2.1]

radio officer on a foreign ship

For the purpose of Part II of Title III of this Act, a person holding at least a first or second class radiotelegraph operator's certificate complying with the provisions of the radio regulations annexed to the International Telecommunication Convention in force. [47 USC 153]

radio officer on a ship of the United States

For the purpose of Part II of Title III of this Act, a person holding at least a first or second class radiotelegraph operator's license as prescribed and issued by the Commission. When such person is employed to operate a radiotelegraph station aboard a ship of the United States, he is also required to be licensed as a "radio officer" in accordance with the Act of May 12, 1948 (46 USC §229a-h). [47 USC 153]

radioprinter operations

Communications by means of a direct printing radiotelegraphy system using any alphanumeric code, within specified bandwidth limitations, which is authorized for use between private coast stations and their associated ship stations on vessels of less than 1600 gross tons. [47 CFR 80.5]

radio regulations

The latest ITU Radio Regulations to which the United States is a party. [47 CFR 13.3, 97.3]

radio service

An administrative subdivision of the field of radio-communication. In an engineering sense, the subdivisions may be made according to the method of operation, as, for example, mobile service and fixed service. In a regulatory sense, the subdivisions may be descriptive of particular groups of licensees, as, for example, the groups of persons licensed under this part. [47 CFR 5.4]

radiosonde

An automatic radio transmitter in the meteorological aids service usually carried on an aircraft, free balloon, kite or parachute, and which transmits meteorological data. (RR) [47 CFR 2.1]

radio station [1]

A station equipped to engage in radio communication or radio transmission of energy. [47 USC 153]

radio station [2]

A separate transmitter or a group of transmitters under simultaneous common control, including the accessory equipment required for carrying on a radiocommunication service. [47 CFR 21.2]

radio station license (see station license)

radio telecommunication services

Communication services provided by the use of radio, including radiotelephone, radiotelegraph, paging and facsimile service. [47 CFR 22.99]

radiotelegram

A telegram, originating in or intended for a mobile station or a mobile earth station transmitted on all or part of its route over the radiocommunication channels of the mobile service or of the mobile-satellite service. (RR) [47 CFR 2.1]

radiotelegraph

As used in this part shall be construed to include types N0N, A1A, A2A, A3C, F1B, F2B, and F3C emission. [47 CFR 23.1]

radiotelegraph auto alarm

Subject to the provisions of Part II of Title III of this Act, an automatic alarm on a ship of the United States receiving apparatus which responds to the radiotelegraph alarm signal and has been approved by the Commission. On a foreign ship, an automatic alarm receiving apparatus which responds to the radiotelegraph alarm signal and has been approved by the government of the country in which the ship is registered: provided, that the United States and the country in which the ship is registered are parties to the same treaty, convention, or agreement prescribing the requirements for such apparatus. Nothing in this Act or in any other provision of law shall be construed to require the recognition of a radiotelegraph auto alarm as complying with Part II of Title III of this Act, on a foreign ship to such part, where the country in which the ship is registered and the United States are not parties to the same treaty, convention, or agreement prescribing the requirements for such apparatus. [47 USC 153]

radiotelegraph service

Transmission of messages from one place to another by means of radio. [47 CFR 22.99]

radiotelemetry

Telemetry by means of radio waves. (RR) [47 CFR 2.1]

radiotelephone

As used in this part, with respect to operation on frequencies below 30 MHZ, means a system of radiocommunication for the transmission of speech or, in some cases, other sounds by means of amplitude modulation including double sideband (A3E), single sideband (R3E, H3E, J3E) or independent sideband (B3E) transmission. [47 CFR 23.1]

radiotelephone call

A telephone call, originating in or intended for a mobile station or a mobile earth station transmitted on all or part of its route over the radiocommunication channels of the mobile service or of the mobile-satellite service. (RR) [47 CFR 2.1]

radio teleprinting

Radio transmissions to a printing telegraphic instrument having a signal-actuated mechanism for automatically printing received messages. [47 CFR 90.7]

radiotelex call

A telex call, originating in or intended for a mobile station or a mobile earth station, transmitted on all or part of its route over the radiocommunication channels of the mobile service or of the mobile-satellite service. (RR) [47 CFR 2.1]

radio transmission of energy (see transmission of energy by radio)

radio waves or Hertzian waves

Electromagnetic waves of frequencies arbitrarily lower than 3,000 GHz, propagated in space without artificial guide. (RR) [47 CFR 2.1]

rate

The tariffed price per unit of service. [47 CFR 61.3]

rate increase

Any change in a tariff which results in an increased rate or charge to any of the filing carrier's customers. [47 CFR 61.3]

rate level change

A tariff change that only affects the actual rate associated with a rate element, and does not affect any tariff regulations or any other wording of tariff language. [47 CFR 61.3]

rated power output

The normal radio frequency power output capability (peak or average power) of a transmitter, under optimum conditions of adjustment and operation, specified by its manufacturer. [47 CFR 21.2, 94.3]

RBDS

Radio broadcast data system [47 CFR 11.33]

R/C

Radio control [47 CFR 90.79]

RCO

Remote communications outlet [47 CFR 87.351]

RCTA

Radio Technical Commission for Aeronautics [47 CFR 76.611]

RCVR

Receiver [47 CFR 2.1045]

readily achievable

Has the meaning given to it by Section 301(9) of the Americans with Disabilities Act of 1990 (42 USC 12181(9). [47 USC 255]

readily visible

That the nameplate or nameplate data must be visible from the outside of the equipment enclosure. It is preferable that it be visible at all times during normal installation or use, but this is not a prerequisite for grant of equipment authorization. [47 CFR 2.925]

rebroadcast [1]

Reception by radio of the programs or other transmissions of a broadcast station, and the simultaneous or subsequent retransmission of such programs or transmissions by a broadcast station. [47 CFR 74.184]

rebroadcast [2]

The reception by radio of the programs or other signals of a radio or television station and the simultaneous or subsequent retransmission of such programs or signals for direct reception by the general public. [47 CFR 74.784]

reconciliation

The comparison of NSEP service information and the resolution of identified discrepancies. [47 CFR 64.1604, 47 CFR Pt. 216, App.]

record

Any item, collection or grouping of information about an individual that is maintained by the Commission, including but not limited to, such individual's education, financial transactions, medical history, and criminal or employment history, and that contains such individual's name, or the identifying number, symbol, or other identifying particular assigned to the individual, such as a finger or voice print or a photograph. [47 CFR 0.551]

record communication

Any transmission of intelligence which is reduced to visual record form at the point of reception. [47 CFR 21.2]

reduced carrier single-sideband emission

A single-sideband emission in which the degree of carrier suppression enables the carrier to be reconstituted and to be used for demodulation. (RR) [47 CFR 2.1]

reduction of service (see discontinuance of service)

reference axis

Normal to the reference plane and passes through the center of the receiver cap (or the center of the hole array, for handset types that do not have receiver caps). [47 CFR 68.316]

reference black level

The level corresponding to the specified maximum excursion of the luminance signal in the black direction. [47 CFR 73.681]

reference frequency [1]

A frequency having a fixed and specified position with respect to the assigned frequency. The displacement of this frequency with respect to the assigned frequency has the same absolute value and sign that the displacement of the characteristic frequency has with respect to the centre of the frequency band occupied by the emission. (RR) [47 CFR 2.1]

reference frequency [2]

A frequency coinciding with or having a fixed and specified relation to the assigned frequency. This frequency does not necessarily correspond to any frequency in an emission. [47 CFR 21.2]

reference month

That month of a season which is used for determining predicted propagation characteristics for the season. The reference month is January, April, July, or October, as related to the season in which it occurs. [47 CFR 73.701]

reference plane

The planar area containing points of the receiver-end of the handset which, in normal handset use, rest against the ear (see Fig 1). [47 CFR 68.316]

reference white level of the luminance signal

The level corresponding to the specified maximum excursion of the luminance signal in the white direction. [47 CFR 73.681]

referral for litigation

Referral to the Department of Justice for appropriate legal proceedings except where the Commission has the statutory authority to handle the litigation itself. [47 CFR 1.1901]

reflecting satellite

A satellite intended to reflect radiocommunication signals. (RR) [47 CFR 2.1]

regional concentration rules

The provisions of §§73.35, 73.240, and 73.636 of Title 47, Code of Federal Regulations (as in effect June 1, 1983), which prohibit any party from directly or indirectly owning, operating, or controlling three broadcast stations in one or several services where any two of such stations are within 100 miles of the third (measured city-to-city), and where there is a primary service contour overlap of any of the stations. [47 USC 310]

registered protective circuitry

Separate, identifiable and discrete electrical circuitry designed to protect the telephone network from harm which is registered in accordance with the rules and regulations in Subpart C of this part. [47 CFR 68.3]

registered terminal equipment

Terminal equipment which is registered in accordance with the rules and regulations in Subpart C of this part. [47 CFR 68.3]

regulations

The body of carrier prescribed rules in a tariff governing the offering of service in that tariff, including rules, practices, classifications, and definitions. [47 CFR 61.3]

relay station

A fixed station used for the reception and retransmission of the signals of another station or stations. [47 CFR 21.2]

remote access line

An access line (*e.g.*, for WATS or TWX service) between a subscriber's premises in one toll rate center and a serving central office located in a different toll rate center. [47 CFR Pt. 36, App.]

remote control [1]

Operation of a station by a qualified operator on duty at a control position from which the transmitter is not visible but which control position is equipped with suitable control and telemetering circuits so that the essential functions that could be performed at the transmitter can also be performed from the control point. [47 CFR 74.901]

remote control [2]

The use of a control operator who indirectly manipulates the operating adjustments in the station through a control link to achieve compliance with the FCC Rules. [47 CFR 97.3]

remote control operation [1]

Operation of a base station by a properly designated person on duty at a control position from which the transmitter is not visible but that position is equipped with suitable controls so that essential functions can be performed therefrom. [47 CFR 74.401]

remote control operation [2]

Operation of a station by a qualified operator on duty at a control position from which the transmitter is not visible but which control position is equipped with suitable control and telemetering circuits so that the essential functions that could be performed at the transmitter can also be performed from the control point. [47 CFR 78.5]

remote line location

A remotely located subscriber line access unit which is normally dependent upon the central processor of the host office for call processing functions. [47 CFR Pt. 36, App.]

remote pickup broadcast base station

A remote pickup broadcast station authorized for operation at a specified location. [47 CFR 74.401]

remote pickup broadcast mobile station

A remote pickup broadcast station authorized for use while in motion or during halts at unspecified locations. (As used in the subpart, mobile stations include hand-carried, pack-carried and other portable transmitters.) [47 CFR 74.401]

remote pickup broadcast stations

Used in this subpart to include both remote pickup broadcast base stations and remote pickup broadcast mobile stations. [47 CFR 74.401]

remote pickup mobile repeater unit

A vehicular receiver-transmitter repeater used to provide extended communications range for a low-power hand-carried or pack-carried transmitter. [47 CFR 74.401]

remote station

A fixed station in a multiple address radio system that transmits one-way to one or more receive central sites, controls a master station, or is controlled, activated, or interrogated by, and may respond to, a master station. [47 CFR 94.3]

remote trunk arrangement (RTA)

Arrangement that permits the extension of TSPS functions to remote locations. [47 CFR Pt. 36, App.]

repeated

With reference to the commission or omission of any act, the commission or omission of such act more than once or, if such commission or omission is continuous, for more than one day. [47 USC 312]

repeatedly
When used with respect to failures to comply, refers to three or more failures during any 7-year period. [47 USC 554]

repeater [1]
A fixed transmitter that retransmits the signals of other stations. [47 CFR 22.99]

repeater [2]
An amateur station that simultaneously retransmits the transmission of another amateur station on a different channel or channels. [47 CFR 97.3]

repeater station
A fixed station established for the automatic retransmission of radiocommunications received from one or more stations and directed to a specified receiver site. [47 CFR 21.2]

representative of the news media
Any person actively gathering news for an entity that is organized and operated to publish or broadcast news to the public. [47 CFR 0.466]

reservation
That amount or quantity of property kept or set apart for a specific use. [47 CFR Pt. 36, App.]

reserved
Kept or set apart for a specific use. [47 CFR Pt. 36, App.]

respondent
A cable television system operator or a utility against whom a complaint is filed. [47 CFR 1.1402]

restoration
The repair or returning to service of one or more telecommunication services that have experienced a service outage or are unusable for any reason, including a damaged or impaired telecommunications facility. Such repair or returning to service may be done by patching, rerouting, substitution of component parts or pathways, and other means, as determined necessary by a service vendor. [47 CFR Pt. 216, App., 47 CFR 64.1604]

restructured service

An offering which represents the modification of a method of charging or provisioning a service; or the introduction of a new method of charging or provisioning that does not result in a net increase in options available to customers. [47 CFR 61.3]

retirement units

As applied to depreciable telecommunications plant, means those items of plant which when removed (with or without replacement) cause the initiation of retirement accounting entries. [47 CFR 32.9000]

return component

Net investment attributable to a particular element or category multiplied by the authorized annual rate of return. [47 CFR 69.2]

re-usable launch vehicle (RLV)

A booster rocket that can be recovered after launch, refurbished and relaunched. [47 CFR 87.5]

revalidation [1]

The rejustification by a service user of a priority level assignment. This may result in extension by the Executive Office of the President of the expiration date associated with the priority level assignment. [47 CFR 64.1604]

revalidation [2]

The rejustification by a service user of a priority level assignment. This may result in extension by the Manager, NCS, in accordance with this directive, of the expiration date associated with the priority level assignment. [47 CFR Pt. 216, App.]

review

The process of examining documents located in response to a commercial use request to determine whether any portion of a document located is exempt form disclosure. It also includes processing any documents for disclosure, *e.g.*, performing such functions that are necessary to excise them or otherwise prepare then for release. Review does not include time spent resolving general legal or policy issues regarding the application of FOIA exemptions. [47 CFR 0.466]

revision

A change in priority level assignment for an NSEP telecommunications service. This includes any extension of an existing priority level assignment to an expanded NSEP service. [47 CFR Pt, 216, App., CFR , CFR 64.1604]

revocation

The elimination of a priority level assignment when it is no longer valid. All priority level assignments for an NSEP service are revoked upon service termination. [47 CFR Pt. 216, App., 47 CFR 64.1604]

RF

Radio frequency [47 CFR 15.3]

right-hand (or clockwise) polarized wave

An elliptically or circularly-polarized wave, in which the electric field vector, observed in any fixed plane, normal to the direction of propagation, whilst looking in the direction of propagation, rotates with time in a right-hand or clockwise direction. (RR) [47 CFR 2.1]

ringdown private line interface

The point of connection between ringdown voiceband private line service and terminal equipment or systems which provide ringing (20 or 30 Hz) in either direction for alerting only. All tip and ring leads shall be treated as telephone connections for the purposes of fulfilling registration conditions. On 2-wire circuits the ringing voltage is applied to the ring conductor with the tip conductor grounded. On 4-wire circuits the ringing voltage is simplexed on the tip and ring conductors with ground simplexed on the tip (1) and ring (1) conductors. [47 CFR 68.3]

RMS

Root-mean-squared [47 CFR 68.308]

RMS

Value of the necessary voltage [47 CFR 2.1041]

RMT

Required monthly test [47 CFR 11.31]

roamer

A mobile station receiving service from a station or system in the Public Mobile Services other than one to which it is a subscriber. [47 CFR 22.99]

routine use

With respect to the disclosure of a record, the use of such record for a purpose which is compatible with the purpose for which it was collected. [47 CFR 0.551]

row

One of 15 horizontal divisions of the screen, extending across the full height of the safe caption area as defined in paragraph (n)(12) of this section. [47 CFR 15.119]

RP

Restoration priority [47 CFR 64.1604]

RPOA

Recognized private operating agency [47 CFR 0.261]

RR

Radio Regulations, Geneva, 1982. [47 CFR 2.1, 25.201]

RSAs

Rural service areas [47 CFR 22.909]

RSL

Received signal power level [47 CFR 21.902]

RSS

Root-sum-square [73.182]

RTA

Remote trunk arrangement [47 CFR 36.123]

RTCM

Radio Technical Commission for Maritime Services [47 CFR 80.1061]

RTTY
Narrow-band direct-printing telegraphy emissions having designators with A, C, D, F, G, H, J or R as the first symbol; 1 as the second symbol; B as the third symbol; and emission J2B. Only a digital code of a type specifically authorized in this part may be transmitted. [47 CFR 97.3]

RTU
Right to use [47 CFR 95.803]

rural area
A community unit with a density of less than 19 households per route kilometer or thirty households per route mile of coaxial and/or fiber optic cable trunk and feeder line. [47 CFR 76.5]

rural radiotelephone service
A radio service in which common carriers are authorized to offer and provide radio telecommunication services for hire to subscribers in areas where it is not feasible to provide communication services by wire or other means. [47 CFR 22.99]

rural subscriber station
One or more fixed transmitters in the Rural Radiotelephone Service that receive service from central office transmitters. [47 CFR 22.99]

rural telephone company [1]
A local carrier operating entity to the extent that such entity--

(A) provides common carrier service to any local exchange carrier study area that does not include either (i) any incorporated place of 10,000 inhabitants or more, or any part thereof, based on the most recently available population statistics of the Bureau of the Census; or (ii) any territory, incorporated or unincorporated, included in an urbanized area, as defined by the Bureau of the Census as of August 10, 1993;

(B) provides telephone exchange service, including exchange access, to fewer than 50,000 access lines;

(C) provides telephone exchange service to any local exchange carrier study area with fewer than 100,000 access lines; or

(D) has less than 15 percent of its access lines in communities of more than 50,000 on the date of enactment of the Telecommunications Act of 1996. [47 USC 153]

rural telephone company [2]

A local exchange carrier having 100,000 or fewer access lines, including all affiliates. [47 CFR 24.4, 26.4, 90.814]

RWT

Required weekly test [47 CFR 11.31]

RX

Receiver [47 CFR 2.1045]

S

SAB
Service area boundary [47 CFR 22.165]

safe caption area
The area of the television picture within which captioning and text shall be displayed to ensure visibility of the information on the majority of home television receivers. The safe caption area is specified as shown in the following figure. [47 CFR 15.119]

safety communication
The transmission or reception of distress, alarm, urgency, or safety signals, or any communication preceded by one of these signals, or any form of radiocommunication which, if delayed in transmission or reception, may adversely affect the safety of life or property. [47 CFR 80.5]

safety convention
The International Convention for the Safety of Life at Sea in force and the regulations referred to therein. [47 USC 153]

safety service
Any radiocommunication service used permanently or temporarily for the safeguarding of human life and property. (CONV) [47 CFR 2.1]

safety signal
(1) The safety signal is the international radiotelegraph or radiotelephone signal which indicates that the station sending this signal is preparing to transmit a message concerning the safety of navigation or giving important meteorological warnings.

(2) In radiotelegraphy, the international safety signals consists of three repetitions of the group "TTT", sent before the call, with the letters of each group and the successive groups clearly separated from each other.

(3) In radiotelephony, the international safety signal consists of three oral repetitions of "Security", pronounced as the French word "Securite", sent before the call. [47 CFR 80.5]

salary offset

An administrative offset to collect a debt under 5 U.S.C. 5514 by deduction(s) at one or more officially established pay intervals from the current pay account of an employee without his or her consent. [47 CFR 1.1901]

salvage value

The amount received for property retired, if sold, or if retained for reuse, the amount at which the material recovered is chargeable to Account 1220, Material and Supplies, or other appropriate account. [47 CFR 32.9000]

SAR

Search and rescue [47 CFR 80.5]

SARDA

State and regional disaster airlift planning [47 CFR 87.393]

satellite

A body which revolves around another body of preponderant mass and which has a motion primarily and permanently determined by the force of attraction of that other body. (RR) [47 CFR 2.1]

satellite broadcast programming

Broadcast video programming when such programming is retransmitted by satellite and the entity retransmitting such programming is not the broadcaster or an entity performing such retransmission on behalf of and with the specific consent of the broadcaster. [47 USC 548, 47 CFR 76.1000]

satellite broadcast programming vendor

A fixed service satellite carrier that provides service pursuant to Section 119 of Title 17, United States Code, with respect to satellite broadcast programming. [47 USC 548, 47 CFR 76.1000]

satellite cable programming [1]

The meaning provided under Section 705 of this Act, except that such term does not include satellite broadcast programming. [47 USC 548]

satellite cable programming [2]

Video programming which is transmitted via satellite and which is primarily intended for the direct receipt by cable operators for their retransmission to cable subscribers. [47 USC 605]

satellite cable programming [3]

Video programming which is transmitted via satellite and which is primarily intended for direct receipt by cable operators for their retransmission to cable subscribers, except that such term does not include satellite broadcast programming. [47 CFR 76.1000]

satellite cable programming vendor

A person engaged in the production, creation, or wholesale distribution for sale satellite cable programming, but does not include a satellite broadcast programming vendor. [47 USC 548, 47 CFR 76.1000]

satellite carrier

Has the meaning given those terms, respectively, in Section 119(d) of title 17, United States Code, as in effect on the date of enactment of the Cable Television Consumer Protection and Competition Act of 1992. [47 USC 325]

satellite link

A radio link between a transmitting earth station and a receiving earth station through one satellite. A satellite link comprises one up-link and one down-link. (RR) [47 CFR 2.1]

satellite network

A satellite system or a part of a satellite system, consisting of only one satellite and the cooperating earth stations. (RR) [47 CFR 2.1]

satellite system

A space system using one or more artificial earth satellites. (RR) [47 CFR 2.1, 25.201]

SBI (see Service Band Index)

SCA

Subsidiary communications authorization [47 CFR 74.531]

scanning

The process of analyzing successively, according to a predetermined method, the light values of picture elements constituting the total picture area. [47 CFR 73.681]

scanning line

A single continuous narrow strip of the picture area containing highlights, shadows, and half-tones, determined by the process of scanning. [47 CFR 73.681]

scanning receiver

A receiver that automatically switches among four or more frequencies in the range of 30 to 960 MHz and which is capable of stopping at and receiving a radio signal detected on a frequency. Receivers designed solely for the reception of the broadcast signals under part 73 of this chapter or for operation as part of a licensed station are not included in this definition. [47 CFR 15.3]

SCATANA

Security control of air traffic and air navigation aids [47 CFR 87.393]

scramble

To rearrange the content of the signal of the programming so that the programming cannot be viewed or heard in an understandable manner. [47 USC 561]

search

All time spent looking for material that is responsive to a request, including page-by-page or line-by-line identification of material contained within documents. Such activity should be distinguished, however, from "review" of material in order to determine whether the material is exempt from disclosure (see paragraph (a)(3) of this section). [47 CFR 0.466]

seasonal schedule

An assignment, for a season, of a frequency or frequencies, and other technical parameters, to be used by a station for transmission to particular zones or areas of reception during specified hours. [47 CFR 73.701]

secondary operation

Radio communications which may not cause interference to operations authorized on a primary basis and which are not protected from interference from those primary operations. [47 CFR 90.7]

secondary radar

A radiodetermination system based on the comparison of reference signals with radio signals retransmitted from the position to be determined. (RR) [47 CFR 2.1]

secondary service area

The service area of a broadcast station served by the skywave and not subject to objectionable interference and in which the signal is subject to intermittent variations in strength. [47 CFR 73.14]

Secretary

The Secretary of Commerce when such term is used in Subpart A and Subpart B, and the Secretary of Health and Human Services when such term is used in Subpart C, Subpart D and this subpart. [47 USC 397]

Section 504

Section 504 of the Rehabilitation Act of 1973, Pub. L. 93-112, 87 Stat. 394, 29 U.S.C. 794, as amended by the Rehabilitation Act Amendments of 1974, Pub. L. 93-516, 88 Stat. 1617, and the Rehabilitation, Comprehensive Services, and Developmental Disabilities Amendments of 1978, Pub. L. 95-602, 92 Stat. 2955, and the Rehabilitation Act Amendments of 1986, sec. 103(d), Pub. L. 99-506, 100 Stat. 1810. As used in this part,

section 504 applies only to programs or activities conducted by Executive agencies and not federally assisted programs.
[47 CFR 1.1803]

secure telephones

Telephones that are approved by the United States Government for the transmission of classified or sensitive voice communications. [47 USC 605, 47 CFR 68.3]

selectable transponder

A transponder whose coded response may be inhibited or displayed on a radar on demand by the operator of that radar. [47 CFR 80.5]

semi-duplex operation

A method which is simplex operation at one end of the circuit and duplex operation at the other. (RR) [47 CFR 2.1]

senior employee

When used in this section, an officer or employee in a position established within the Senior Executive Service (pursuant to the Civil Service Reform Act of 1978) or a position for which the basic rate of pay is equal to or greater than the basic pay rate of GS-17 of the General Schedule prescribed by 5 U.S.C. 5332, and who has significant decisionmaking or supervisory responsibility. Such officers and employees are so designated by the Office of Government Ethics in Consultation with the head of the agency. See 5 CFR 737.25 and 737.33. [47 CFR 1.25]

separated affiliate

A corporation under common ownership or control with a Bell operating company that does not own or control a Bell operating company and is not owned or controlled by a Bell operating company and that engages in the provision of electronic publishing which is disseminated by means of such Bell operating company's or any of its affiliates' basic telephone service. [47 USC 274]

separations

The process by which telecommunication property costs, revenues, expenses, taxes and reserves are apportioned among the operations. [47 CFR Pt. 36, App.]

September season

That portion of any year commencing at 0100 g.m.t. on the first Sunday in September and ending at 0100 g.m.t. on the first Sunday in November. [47 CFR 73.701]

SERS

Special emergency radio service [47 CFR 90.33]

SCATANA

Security control of air traffic and air navigation aids [47 CFR 87.393]

service area [1]

A geographic area established by a state commission for the purpose of determining universal service obligations and support mechanisms. In the case of an area served by a rural telephone company, "service area" means such company's "study area" unless and until the Commission and the states, after taking into account recommendations of a Federal-State Joint Board instituted under section 410(c), establish a different definition of service area for such company." [47 USC 102, 47 USC 214]

service area [2]

The geographic area considered by the FCC to be reliably served by a station in the Public Mobile Services. [47 CFR 22.99]

Service Band Index (SBI)

An index of the level of aggregate rate element rates in a service category, which index is calculated pursuant to §61.47. [47 CFR 61.3]

service category

Any group of rate elements subject to price cap regulation, which group is subject to a band. [47 CFR 61.3]

service contour

The locus of points surrounding a transmitter where the predicted median field strength of the signal from that transmitter is the minimum field strength that is considered sufficient to provide reliable service to mobile stations. [47 CFR 22.99]

service identification

Information uniquely identifying an NSEP telecommunications service to the service vendor and/or service user. [47 CFR Pt., 216, App., 47 CFR 64.1604]

service interruption

The loss of picture or sound on one or more cable channels. [47 CFR 76.309]

service observing unit

A unit of work measurement which is used as the common denominator to express the relative time required for handling the various work functions at service observing boards. [47 CFR Pt. 36, App.]

service or other valuable consideration

Shall not include any service or property furnished without charge or at a nominal charge for use on, or in connection with, a broadcast, or for use on a program which is intended for broadcasting over any radio station, unless it is so furnished in consideration for an identification in such broadcast or in such program of any person, product, service, trademark, or brand name beyond an identification which is reasonably related to the use of such service or property in such broadcast or such program. [47 USC 507]

service tier

A category of cable service or other services provided by a cable operator and for which a separate rate is charged by the cable operator. [47 USC 522]

service to subscribers

Service to at least one subscriber that is not affiliated with, controlled by or related to the providing carrier. [47 CFR 22.99]

service user

Any individual or organization (including a service vendor) supported by a telecommunications service for which a priority level has been requested or assigned pursuant to section 8 or 9 of this appendix. [47 CFR Pt. 216 App., 47 CFR 64.1604]

service vendor

Any person, association, partnership, corporation, organization, or other entity (including common carriers and government organizations) that offers to supply any telecommunications equipment, facilities, or services (including customer premises equipment and wiring) or combination thereof. The term includes resale carriers, prime contractors, subcontractors, and interconnecting carriers. [47 CFR Pt. 216, App., 47 CFR 64.1604]

services (see spare circuits)

serving wire center

The telephone company central office designated by the telephone company to serve the geographic area in which the interexchange carrier or other person's point of demarcation is located. [47 CFR 69.2]

share of viewing hours

The total hours that noncable television households viewed the subject station during the week, expressed as a percentage of the total hours these households viewed all stations during the period. [47 CFR 76.5]

sheath kilometers

The actual length of cable in route kilometers. [47 CFR Pt. 36, App.]

SHF

Super-high frequency [47 CFR 97.3]

ship (vessel) [1]

Every description of watercraft or other artificial contrivance, except aircraft, used or capable of being used as a means of transportation on water, whether or not it is actually afloat. [47 USC 153]

ship or vessel [2]

Includes every description of watercraft or other artificial contrivance, except aircraft, capable of being used as a means of transportation on water whether or not it is actually afloat. [47 CFR 80.5]

ship earth station

A mobile earth station in the maritime mobile-satellite service located on board ship. (RR) [47 CFR 2.1, 80.5]

ship movement service

A safety service in the maritime mobile service other than a port operations service, between coast stations and ship stations, or between ship stations, in which messages are restricted to those relating to the movement of ships. Messages which are of a public correspondence nature shall be excluded from this service. (RR) [47 CFR 2.1]

ship radio station license

An authorization issued by the Commission to operate a radio station onboard a vessel. [47 CFR 80.5]

ship station [1]

Has the meaning given it by the Commission by rule. [47 USC 403]

ship station [2]

A mobile station in the maritime mobile service located onboard a vessel which is not permanently moored, other than a survival craft station. (RR) [47 CFR 2.1, 80.5]

ship's emergency transmitter

A ship's transmitter to be used exclusively on a distress frequency for distress, urgency or safety purposes. (RR) [47 CFR 2.1]

SHL

Studio to headend link [47 CFR 78.5]

short haul system

A microwave system licensed under this part in which the longest radio circuit of tandem radio paths does not exceed 402 km (250 miles). [47 CFR 94.3]

SIDs

System identification numbers [47 CFR 22.941]

signal booster

A device which automatically amplifies and transmits received base station transmissions with no change in frequency or authorized bandwidth. [47 CFR 90.7]

signal booster station

A low-power repeater station automatically retransmitting on the same frequency as the received signal, and located within the protected service area of a Multipoint Distribution Service Station. [47 CFR 21.2]

signaling system 7 (SS7)

A carrier to carrier out-of-band signaling network used for call routing, billing and management. [47 CFR 64.1600]

signalling for tandem switching

The carrier identification code (CIC) and OZZ code, or equivalent information needed to perform tandem switching functions. The CIC identifies the interexchange carrier and the OZZ identifies the interexchange carrier trunk to which traffic should be routed. [47 CFR 69.2]

signed

As used in this section, an original hand-written signature, except that by public notice in the *FEDERAL REGISTER* the Common Carrier Bureau may allow signature by any symbol executed or adopted by the applicant with the intent that such symbol be a signature, including symbols formed by computer-generated electronics impulses. [47 CFR 1.743]

significant interest

A cognizable interest for attributing interests in broadcast, cable, and newspaper properties pursuant to §§73.3555, 73.3615, and 76.501. [47 CFR 76.5]

significantly viewed

Viewed in other than cable television households as follows: (1) For a full or partial network station—a share of viewing hours of at least 3 percent (total week hours), and a net weekly circulation of at least 25 percent; and (2) for an independent station—a share of viewing hours of at least 2 percent (total week hours), and a net weekly circulation of at least 5 percent. See §76.54. [47 CFR 76.5]

similarly situated

For the purposes of evaluating alternative programming contracts offered by a defendant programming vendor, that an alternative multichannel video programming distributor has been identified by the defendant as being more properly compared to the complainant in order to determine whether a violation of §76.1002(b) has occurred. The analysis of whether an alternative multichannel video programming distributor is properly comparable to the complainant includes consideration of, but is not limited to, such factors as whether the alternative multichannel video programming distributor operates within a geographic region proximate to the complainant, has roughly the same number of subscribers as the complainant, and purchases a similar service as the complainant. Such alternative multichannel video programming distributor, however, must use the same distribution technology as the "competing" distributor with whom the complainant seeks to compare itself. [47 CFR 76.1000]

simplex operation

Operating method in which transmission is made possible alternatively in each direction of a telecommunication channel, for example, by means of manual control. (RR) [47 CFR 2.1]

single-sideband emission

An amplitude modulated emission with one sideband only. (RR) [47 CFR 2.1]

single unit installations

For single unit installations existing as of August 13, 1990, and installations installed after that date the demarcation pint shall be a point within 30 cm (12 in) of the protector or, where there is no protector, within 30 cm (12 in) of where the telephone wire enters the customer's premises. [47 CFR 68.3]

SLU

Subscriber line use [47 CFR 36.154]

small base station

Any base station that (1) has an antenna no more than 6.1 meters (20 feet) above the ground or above the building or tree on which it is mounted (see §95.51); and (2) transmits with no more than 5 watts ERP. [47 CFR 95.25]

small business [1]

An entity that together with its affiliates has average annual gross revenues that are not more than $40 million for the preceding three calendar years. [47 CFR 21.961]

small business [2]

An entity that:

(i) Together with its affiliates has average annual gross revenues that are not more than $40 million for the preceding three calendar years;

(ii) Has no attributable investor or affiliate that has a personal net worth of $40 million or more;

(iii) Has a control group all of whose members are affiliates are considered in determining whether the entity meets the $40 million annual gross revenues and personal net work standards; and

(iv) Such control group holds 50.1 percent of the entity's voting interest, if a corporation, and at least 25 percent of the entity's equity on a fully diluted basis, except that a business owned by members of minority groups and/or women (as defined in paragraph (c) of this section) may also qualify as a small business if a control group that is 100 percent composed of members of minority groups and/or women holds 50.1 percent of the entity's voting interests, if a corporation, and 50.1 percent of the entity's total equity on a fully diluted basis and no single other investor holds more than 49.9 percent of passive equity in the entity. [47 CFR 24.320]

small business [3]

(1) An entity that, together with its affiliates and persons or entities that hold interest in such entity and their affiliates, has average annual gross revenues that are not more than $40 million for the preceding three years.

(2) For purposes of determining whether an entity meets the $40 million average annual gross revenues size standard set forth in paragraph (b)(1) of this section, the gross revenues of the entity, its affiliates, persons or entities holding interests in the entity and their affiliates shall be considered on a cumulative basis and aggregated, subject to the exceptions set forth §§24.709(b) or 24.715(b). [47 CFR 24.720]

small business [4]

An individual or business entity which, at the time of application to the Commission, had, including all affiliated entities under common control, annual revenues of less than $500,000 and assets of less than $1,000,000. [47 CFR 73.3555]

small business consortium [1]

A conglomerate organization formed as a joint venture between mutually-independent business firms, each of which individually satisfies the definition of a small business. [47 CFR 21.961]

small business consortium [2]

A conglomerate organization formed as a joint venture between or among mutually-independent business firms, each of which individually satisfies the definition of a small business in paragraphs (b)(1) and (b)(2) of this section. [47 CFR 24.230, 24.720]

small business: consortium of small businesses [1]

(1) A small business is an entity that, together with its affiliates and persons or entities that hold interest in such entity and their affiliates, has average annual gross revenues that are not more than $40 million for the preceding three years.

(2) A small business consortium is a conglomerate organization formed as a joint venture between or among mutually-independent business firms, each of which individually satisfies the definition of a small business. [47 CFR 26.4]

small business: consortium of small business [2]

(1) A small business is an entity that either (i) together with its affiliates, persons or entities that hold attributable interests in such entity, and their affiliates, has average gross revenues that are not more than $3 million for the preceding three years; or (ii) together with its affiliates, persons or entities that hold attributable interests in such entity, and their affiliates, has average gross revenues that are not more than $15 million for the preceding three years.

(2) For purposes of determining whether an entity meets either the $3 million or $15 million average annual gross revenues size standard set forth in paragraph (b)(1) of this section, the gross revenues of the entity, its affiliates, persons or entities holding interests in the entity and their affiliates shall be considered on a cumulative basis and aggregated, subject to the exceptions set forth in §90.814(g).

(3) A small business consortium is a conglomerate organization formed as a joint venture between or among mutually-independent business firms, each of which individually satisfies either definition of a small business in paragraphs (b) (1) and (b) (2) of this section. In a consortium of small businesses, each individual member must establish its eligibility as a small business, as defined in this section. [47 CFR 90.814]

small business owned by members of minority groups and/or women
An entity that meets the definitions in both paragraphs (b) and (c) of this section. [47 CFR 24.720]

small cable operator
A cable operator that, directly or through an affiliate, serves in the aggregate fewer than 1 percent of all subscribers in the United States and is not affiliated with any entity or entities whose gross annual revenues in the aggregate exceed $250,000,000. [47 USC 301]

small control station
Any control station which (1) has an antenna no more than 6.1 meters (20 feet) above the ground or above the building or tree on which it is mounted (see §95.51); and (2) is: (i) South of Line A or west of Line C (see §95.37); or (ii) North of Line A or east of Line C. And the station transmits with no more than 5 watts ERP (effective radiated power). [47 CFR 95.25]

smaller television market
The specified zone of a commercial television station licensed to a community that is not listed in §76.51. [47 CFR 76.5]

SMATV
Satellite master antenna television service [47 CFR 76.501]

SMR
Specialized mobile radio [47 CFR 90.617]

SMRS
Specialized mobile radio system [47 CFR 1.925]

SMSA (standard metropolitan statistical area)
A city of 50,000 or more population and the surrounding counties. [47 CFR 90.7]

SP
State primary [47 CFR 11.18]

spacecraft
A man-made vehicle which is intended to go beyond the major portion of the earth's atmosphere. (RR) [47 CFR 2.1, 25.201]

space operation service
A radiocommunication service concerned exclusively with the operation of spacecraft, in particular space tracking, space telemetry, and space telecommand. Note: These functions will normally be provided within the service in which the space station is operating. (RR) [47 CFR 2.1, 25.201]

space radiocommunication
Any radiocommunication involving the use of one or more space stations or the use of one or more reflecting satellites or other objects in space. (RR) [47 CFR 2.1, 25.201]

space research service
A radiocommunication service in which spacecraft or other objects in space are used for scientific or technological research purposes. (RR) [47 CFR 2.1]

space station [1]
A station located on an object which is beyond, is intended to go beyond, or has been beyond, the major portion of the earth's atmosphere. (RR) [47 CFR 2.1, 25.201]

space station [2]
An amateur station located more than 50 km above the earth's surface. [47 CFR 97.3]

space system
Any group of cooperating earth stations and/or space stations employing space radiocommunication for specific purposes. (RR) [47 CFR 2.1, 25.201]

space telecommand
The use of radiocommunication for the transmission of signals to a space station to initiate, modify or terminate functions of equipment on a space object, including the space station. (RR) [47 CFR 2.1, 25.201]

space telemetering
The use of telemetering for the transmission from a space station of results of measurements made in a spacecraft, including those relating to the functioning of the spacecraft. [47 CFR 25.201]

space telemetry [1]
The use of telemetry for transmission for a space station of results of measurements made in a spacecraft, including those relating to the functioning of the spacecraft. (RR) [47 CFR 2.1]

space telemetry [2]
A one-way transmission from a space station of measurements made from the measuring instruments in a spacecraft, including those relating to the functioning of the spacecraft. [47 CFR 97.3]

space tracking
Determination of the orbit, velocity or instantaneous position of an object in space by means of radiodetermination, excluding primary radar, for the purpose of following the movement of the object. (RR) [47 CFR 2.1, 25.201]

spare circuits (services) [1]
Those not being used or contracted for by any customer. [47 CFR 64.1604]

"spare" circuits or services [2]
Circuits or services not being used or contracted for by any customer. [47 CFR Pt. 216, App.]

special characters

Displayable characters (except for "transparent space") which require a two-byte sequence of one non-printing and one printing character. The non-printing byte varies depending on the data channel. Regular characters require unique one-byte codes which are the same in either data channel. [47 CFR 15.119]

special government employee

A special government employee, as defined in section 202 of title 18 of the United States Code, that is, one appointed to serve with or without compensation, for not more than 130 days during any period of 365 days on a full-time or intermittent basis, who is employed in the Commission, but does not include a member of the uniformed services. [47 CFR 19.735-102]

special offices and facilities

Major airports, major military installations, key government facilities, nuclear power plants and 911 public service answering points (PSAPs). [47 CFR 63.100]

special service

A radiocommunication service, not otherwise defined in this Section, carried on exclusively for specific needs of general utility, and not open to public correspondence. (RR) [47 CFR 2.1]

special services

All services other than message telephones, *e.g.*, teletypewriter exchange service (TWX), private line services. [47 CFR Pt. 36, App.]

specialized mobile radio system

A radio system in which licensees provide land mobile communications services (other than radiolocation services) in the 800 MHZ and 900 MHZ bands on a commercial basis to entities eligible to be licensed under this Part, federal government entities, and individuals. [47 CFR 90.7]

specialty adapters

Adapters that contain passive components such as resistive pads or bias resistors typically used for connecting data equipment having fixed-low loop or programmed data jack network connections to key systems or PBXs. [47 CFR 68.3]

specified zone of A television broadcast station

The area extending 56.3 air km (35 air miles) from the reference point in the community to which that station is licensed or authorized by the Commission. A list of reference points is contained in §76.53. A television broadcast station that is authorized but not operating has a specified zone that terminates eighteen (18) months after the initial grant of its construction permit. [47 CFR 76.5]

spectrum window

An amount of spectrum equal to the intended emission bandwidth in which operation is desired. [47 CFR 15.303]

spousal affiliation

Both spouses are deemed to own or control or have the power to control interests owned or controlled by either of them, unless they are subject to a legal separation recognized by a court of competent jurisdiction in the United States. [47 CFR 24.720, 47 CFR 90.814]

spread spectrum systems

A spread spectrum system is an information bearing communications system in which: (1) Information is conveyed by modulation of a carrier by some conventional means, (2) the bandwidth is deliberately widened by means of a spreading function over that which would be needed to transmit the information alone. (In some spread spectrum systems, a portion of the information being conveyed by the system may be contained in the spreading function.) [47 CFR 2.1]

SPS

Special weather statement [47 CFR 11.31]

spurious emission [1]

Emission on a frequency or frequencies which are outside the necessary bandwidth and the level of which may be reduced without affecting the corresponding transmission of information. Spurious emissions include harmonic emissions, parasitic emissions, intermodulation products and frequency conversion products, but exclude out-of-band emissions. (RR) [47 CFR 2.1]

spurious emission [2]

An emission, or frequencies outside the necessary bandwidth of a transmission, the level of which may be reduced without affecting the information being transmitted. [47 CFR 97.3]

SR

State relay [47 CFR 11.18]

SRC

Subsidiary record categories [47 CFR 36.302]

SS7

Signaling system 7 [47 CFR 64.1600]

STA

Special temporary authorization [47 CFR 1.1707]

standard frequency and time signal station [1]

A station in the standard frequency and time signal service. (RR) [47 CFR 2.1]

standard frequency and time signal service [2]

A radiocommunication service for scientific, technical and other purposes, providing the transmission of specified frequencies, time signals, or both, of stated high precision, intended for general reception. (RR) [47 CFR 2.1]

standard frequency and time signal-satellite service

A radiocommunication service using space stations on earth satellites for the same purposes as those of the standard frequency and time signal service. Note: This service may also include feeder links necessary for its operation. (RR) [47 CFR 2.1]

standard television signal

A signal which conforms to the television transmission standards. [47 CFR 73.681]

standby transmitter

A transmitter installed and maintained for use in lieu of the main transmitter only during periods when the main transmitter is out of service for maintenance or repair. [47 CFR 21.2]

state [1]

Any of the several states, the District of Columbia, or any territory or possession of the United States. [47 USC 153, 47 USC 602]

state [2]

Includes the District of Columbia, the Commonwealth of Puerto Rico, the Virgin Islands, Guam, American Samoa, the Northern Mariana Islands, and the Trust Territory of the Pacific Islands. [47 USC 397]

state [3]

Any state, territory, or possession of the United States, the District of Columbia, or any political subdivision, agency, or instrumentality thereof. [47 CFR 1.1402]

state commission

The commission, board, or official (by whatever name designated) which under the laws of any state has regulatory jurisdiction with respect to intrastate operations of carriers. [47 USC 153]

station [1]

One or more transmitters or receivers or a combination of transmitters and receivers, including the accessory equipment, necessary at one location for carrying on a radiocommunication service, or the radio astronomy service. Note: Each station shall be classified by the service in which it operates permanently or temporarily. (RR) [47 CFR 2.1, 80.5]

station [2]

A station equipped to engage in radio communication or radio transmission of energy (47 U.S.C. 153(k)). [47 CFR 22.99]

station [3]

As used in this subpart, each remote pickup broadcast transmitter, and its associated accessory equipment necessary to the radio communication function, constitutes a separate station. [47 CFR 74.401]

station authorization [1]

Any construction permit, license, or special temporary authorization issued by the Commission. [47 CFR 5.4]

station authorization [2]

A license issued by the Commission for the operation of a radio station. [47 CFR 90.7]

station files

Applications, notifications, correspondence and other materials, and copies of authorizations, comprising technical, legal, and administrative data relating to each station in the Public Mobile Services are maintained by the FCC in individual station files. These files constitute the official records for these stations and supersede any other records, data bases or lists from the FCC or other sources. [47 CFR 22.101]

station license (radio station license or license)

That instrument of authorization required by this Act or the rules and regulations of the commission made pursuant to this Act, for the use or operation of apparatus for transmission of energy, or communications, or signals by radio, by whatever name the instrument may be designated by the Commission. [47 USC 153]

station licensee (see licensee)

station-to-station basis

The term applied to the basis of toll rate making which contemplates that the message toll service charge (telephone or TWX) covers the use made of all facilities between the originating station and the terminating station, including the stations, and the services rendered in connection therewith. [47 CFR Pt. 36, App.]

stereophonic channel

The band of audio frequencies from 50 to 10,000 Hz containing the stereophonic information which modulates the radio frequency carrier. [47 CFR 73.14]

stereophonic crosstalk

An undesired signal occurring in the main channel from modulation of the stereophonic channel or that occurring in the stereophonic channel from modulation of the main channel. [47 CFR 73.14]

stereophonic pilot tone

An audio tone of fixed or variable frequency modulating the carrier during the transmission of stereophonic programs. [47 CFR 73.14]

stereophonic separation [1]

The ratio of the electrical signal caused in the right (or left) stereophonic channel to the electrical signal caused in the left (or right) stereophonic channel by the transmission of only a right (or left) signal. [47 CFR 73.14]

stereophonic separation [2]

The ratio of the electrical signal caused in sound channel A to the signal caused in sound channel B by the transmission of only a channel B signal. Channels A and B may be any two channels of a stereophonic sound broadcast transmission system. [47 CFR 73.310]

stereophonic sound

The audio information carried by plurality of channels arranged to afford the listener a sense of the spatial distribution of sound sources. Stereophonic sound broadcasting includes, but is not limited to, biphonic (two channel), triphonic (three channel) and quadrophonic (four channel) program services. [47 CFR 73.310]

stereophonic sound broadcasting

Cross-talk. An undesired signal occurring in one channel caused by an electrical signal in another channel.

FM stereophonic broadcast. The transmission of a stereophonic program by a single FM broadcast station utilizing the main channel and a stereophonic subchannel.

Left (or right) signal. The electrical output of a microphone or combination of microphones placed so as to convey the intensity, time, and location of sounds originating predominately to the listener's left (or right) of the center of the performing area.

Left (or right) stereophonic channel. The left (or right) signal as electrically reproduced in reception of FM stereophonic broadcasts.

Main channel. The band of frequencies from 50 to 15,000 Hz which frequency-modulate the main carrier.

Pilot subcarrier. A subcarrier that serves as a control signal for use in the reception of FM stereophonic sound broadcasts.

Stereophonic separation. The ratio of the electrical signal caused in sound channel A to the signal caused in sound channel B by the transmission of only a channel B signal. Channels A and B may be any two channels of a stereophonic sound broadcast transmission system.

Stereophonic sound. The audio information carried by plurality of channels arranged to afford the listener a sense of the spatial distribution of sound sources. Stereophonic sound broadcasting includes, but is not limited to, biphonic (two channel), triphonic (three channel) and quadrophonic (four channel) program services.

Stereophonic sound subcarrier. A subcarrier within the FM broadcast baseband used for transmitting signals for stereophonic sound reception of the main broadcast program service.

Stereophonic sound subchannel. The band of frequencies from 23 kHz to 99 kHz containing sound subcarriers and their associated sidebands. [47 CFR 73.310]

stereophonic sound subcarrier

A subcarrier within the FM broadcast baseband used for transmitting signals for stereophonic sound reception of the main broadcast program service. [47 CFR 73.310]

stereophonic sound subchannel

The band of frequencies from 23 kHz to 99 kHz containing sound subcarriers and their associated sidebands. [47 CFR 73.310]

STL

Studio transmitter link [47 CFR 21.2]

STP

Signalling transfer point [47 CFR 69.125]

straight-line method

As applied to depreciation accounting, means the plan under which the cost of property is charged to operating expenses and credited to accumulated depreciation through equal annual charges as nearly as may be during its service life. [47 CFR 32.9000]

studio

Any room or series of rooms equipped for the regular production of broadcast programs of various kinds. A broadcasting booth at a stadium, convention hall, church, or other similar place is not considered to be a studio. [47 CFR 74.401]

studio to transmitter link (STL)

A directional path used to transmit a signal from a station's studio to its transmitter. [47 CFR 74.901]

study area

Study area boundaries shall be frozen as they are on November 15, 1984. [47 CFR Pt. 36, App.]

sub-band

For purposes of this subpart the term sub-band refers to the spectrum allocated for isochronous or asynchronous transmission. [47 CFR 15.303]

subdistribution agreement

An arrangement by which a local cable operator is given the right by a satellite cable programming vendor or satellite broadcast programming vendor to distribute the vendor's programming to competing multichannel video programming distributors. [47 CFR 76.1000]

subrate digital service

A digital service providing for the full-time simultaneous two-way transmission of digital signals at synchronous speeds of 2.4, 4.8, 9.6 or 56 kbps. [47 CFR 68.3]

subscriber line or exchange line

A communication channel between a telephone station, PBX or TWX station and the central office which serves it. [47 CFR Pt. 36, App.]

subscriber list information

Any information (A) identifying the listed names of subscribers of a carrier and such subscribers' telephone numbers, addresses, or primary advertising classifications (as such classifications are assigned at the time of the establishment of such service), or any combination of such listed names, numbers, addresses, or classifications; and (B) that the carrier or an affiliate has published, caused to be published, or accepted for publication any directory format. [47 USC 222]

subscriber line cable and wire facilities

All lines or trunks on the subscriber side of a Class 5 or end office switch, including lines or trunks that do not terminate in such a switch, except lines or trunks that connect an interexchange carrier. [47 CFR 69.2]

subscriber terminal

The cable television system terminal to which a subscriber's equipment is connected. Separate terminals may be provided for delivery of signals of various classes. [47 CFR 76.5]

subscribers

A member of the general public who receives broadcast programming distributed by a cable television system and does not further distribute it. [47 CFR 76.5]

subsidiary channel

A subsidiary channel is any portion of an authorized channel not used for main channel transmissions. [47 CFR 74.901]

subsidiary record

Accumulation of detailed information which is required by this Commission to be maintained in support of entries to the accounts. [47 CFR 32.9000]

subsidiary record categories

Those segregations of certain regulated costs, expenses and revenues which must be maintained and are subject to specific reporting requirements of this Commission. [47 CFR 32.9000]

subsystems, parallel mechanisms

Processes or procedures which augment the use of a chart of accounts in the financial operation of the entity. These subsystems operate on and/or process account and subsidiary record information for specific purposes. [47 CFR 32.9000]

subtributary office

A class of tributary office which does not have direct access to its toll center, but which is connected to its toll center office by means of circuits which are switched through to the toll center at another tributary office. [47 CFR Pt. 36, App.]

sunrise and sunset

For each particular location and during any particular month, the time of sunrise and sunset as specified in the instrument of authorization (See §73.1209). [47 CFR 73.14]

sunshine agenda period

For purposes of this subpart, the Sunshine Agenda period is defined as the period of time that commences with release of a public notice that a matter has been placed on the Sunshine Agenda and that terminates when the Commission (1) releases the text of a decision or order relating to the matter, or (2) issues a public notice stating that the matter has been deleted from the Sunshine Agenda, or (3) issues a public notice stating that the matter has been returned to the staff for further consideration, whichever one of the above occurs first. [47 CFR 1.1202]

sunspot number

The 12-month running average of the number of sunspots for any month as indicated in the U.S. Department of Commerce Telecommunications Research and Engineering Report No. 13—available from the Superintendent of Documents, Washington, DC 20402. The sunspot number varies in an approximate 11-year cycle. [47 CFR 73.701]

super-high frequency (SHF)

The frequency range 3-30 GHz. [47 CFR 97.3]

superstation

Has the meaning given those terms, respectively, in Section 119(d) of title 17, United States Code, as in effect on the date of enactment of the Cable Television Consumer Protection and Competition Act of 1992. [47 USC 325]

supplement

A publication filed as part of a tariff for the purpose of suspending or cancelling that tariff, or tariff publication and numbered independently from the tariff page series. [47 CFR 61.3]

suppressed carrier single-sideband emission

A single-sideband emission in which the carrier is virtually suppressed and not intended to be used for demodulation. (RR) [47 CFR 2.1]

surveillance radar station

A radionavigation land station in the aeronautical radionavigation service employing radar to display the presence of aircraft within its range. [47 CFR 87.5]

survival craft station

A mobile station in the maritime or aeronautical mobile service intended solely for survival purposes and located on any lifeboat, life raft or other survival equipment. [47 CFR 2.1, 80.5, 87.5]

SVA

Severe thunderstorm watch [47 CFR 11.31]

SVR

Severe thunderstorm warning [47 CFR 11.31]

SVS

Severe weather statement [47 CFR 11.31]

symbol rate

Modulation rate in bands. This rate may be higher than the transmitted bit rate as in the case of coded pulses or lower as in the case of multilevel transmission. [47 CFR 21.2]

synchronization

The maintenance of one operation in step with another. [47 CFR 73.681]

syndicated program

Is any program sold, licensed, distributed or offered to television station licensees in more than one market within the United States other than as network programming as defined in §76.5(o). [47 CFR 76.5]

system community unit: community unit

A cable television system, or portion of a cable television system, that operates or will operate within a separate and distance community or municipal entity (including unincorporated communities within unincorporated areas and including single, discrete unincorporated areas). [47 CFR 76.5]

system manager

The commission official responsible for the storage, maintenance, safekeeping, and disposal of a system of records. [47 CFR 0.551]

system noise

That combination of undesired and fluctuating disturbances within a cable television channel that degrades the transmission of the desired signal and that is due to modulation processes or thermal or other noise-producing effects, but does not include hum and other undesired signals of discrete frequency. System noise is specified in terms of its rms voltage or its mean power level as measured in the 4 MHZ bandwidth between 1.25 and 5.25 MHZ above the lower channel boundary of a cable television channel. [47 CFR 76.5]

system of public telecommunications entities

Any combination of public telecommunications entities acting cooperatively to produce, acquire, or distribute programs, or to undertake related activities. [47 USC 397]

system of records

A group of records under the control of the commission from which information is retrievable by the name of the individual or by some identifying number, symbol, or other identifying particular assigned to the individual. [47 CFR 0.551]

system premises wiring

Wiring which connects separately-housed equipment entities or system components to one another, or wiring which connects an equipment entity or system component with the telephone network interface, located at the customer's premises and not within an equipment housing. [47 CFR 68.3]

systems

A complete remote pickup broadcast facility consisting of one or more mobile stations and/or one or more base stations authorized pursuant to a single license. [47 CFR 74.401]

T

T(OPS)

Telephone off-premises station [47 CFR 68.3]

tandem area

The general areas served by the local offices having direct trunks to or from the tandem office. This area may consist of one or more communities or may include only a portion of a relatively large city. [47 CFR Pt. 36, App.]

tandem circuit or trunk

A general classification of circuits or trunks between a tandem central office unit and any other central office or switchboard. [47 CFR Pt. 36, App.]

tandem connection

A call switched at a tandem office. [47 CFR Pt. 36, App.]

tandem office

A central office unit used primarily as an intermediate switching point for traffic between local central offices within the tandem area. Where qualified by a modifying expression, or other explanation, this term may be applied to an office employed for both the interconnection of local central offices within the tandem area and for the interconnection of these local offices with other central offices, *e.g.*, long haul tandem office. [47 CFR Pt. 36, App.]

tandem-switched transport

Transport of traffic that is switched at a tandem switch (1) between the serving wire center and the end office, or (2) between the telephone company office containing the tandem switching equipment, as described in §36.124 of this chapter, and the end office.

Tandem-switched transport between a serving wire center and an end office consists of circuits dedicated to the use of single interexchange carrier or other person from the serving wire center to the tandem (although this dedicated link will not exist if the serving wire center and the tandem are located in the same place) and circuits used in common by multiple interexchange carriers or other persons from the tandem to the end office. [47 CFR 69.2]

tangible property

Assets characterized by physical existence, such as land, buildings, equipment, furniture, fixtures and tools. [47 CFR 32.9000]

tariff

Schedules of rates and regulations filed by common carriers. [47 CFR 61.3]

tariff publication (publication)

A tariff, supplement, revised page, additional page, concurrence, notice of revocation, adoption notice, or any other schedule of rates or regulations. [47 CFR 61.3]

tariff year

The period from the day in a calendar year on which a carrier's annual access tariff filing is scheduled to become effective through the preceding day of the subsequent calendar year. [47 CFR 61.3]

tax (fee)

Any local sales tax, local use tax, local intangible tax, local income tax, business license tax, utility tax, privilege tax, gross receipts tax, excise tax, franchise fees, local telecommunications tax, or any other tax, license, or fee that is imposed for the privilege of doing business, regulating, or raising revenue for a local taxing jurisdiction. [47 USC 602]

TDD

A Telecommunications Device for the Deaf, which is a machine that employs graphic communication in the transmission of coded signals through a wire or radio communication system. [47 USC 225]

TDDRA

Telephone Disclosure and Dispute Resolution Act [47 CFR 64.1502]

technically integrated

   Having 75% or more of the video channels received from a common headend. [47 CFR 76.5]

telecast

   (A) to broadcast by a television broadcast; or (B) to transmit by a cable television system or a satellite television distribution service. [47 USC 303c]

telecommand [1]

   The use of telecommunication for the transmission of signals to initiate, modify or terminate functions of equipment at a distance. (RR) [47 CFR 2.1]

telecommand [2]

   The transmission of nonvoice signals for the purpose of remotely controlling a device.   [47 CFR 90.7]

telecommand [3]

   A one-way transmission to initiate, modify, or terminate functions of a device at a distance. [47 CFR 97.3]

telecommand station

   An amateur station that transmits communications to initiate, modify or terminate functions of a space station. [47 CFR 97.3]

telecommunications [1]

   The transmission, between or among points specified by the user, of information of the user's choosing, without change in the form or content of the information as sent and received. [47 USC 153]

telecommunications [2]

   Any transmission, emission or reception of signs, signals, writing, images and sounds or intelligence of any nature by wire, radio, optical or other electromagnetic systems. (CONV) [47 CFR 2.1]

telecommunications [3]

   Any transmission, emission, or reception of signs, signals, writing, images or sounds or intelligence of any nature by wire, radio, visual or other electromagnetic systems. This encompasses the aggregate of several modes of conveying information, signals or messages over a distance. Included in the telecommunications industry is the transmitting, receiving, or exchanging of information among multiple locations. The minimum elements required for the telecommunications process to occur are a message source, a transmission medium and a receiver. [47 CFR 32.9000]

telecommunications carrier

Any provider of telecommunications services, except that such term does not include aggregators of telecommunications services (as defined in Section 226). A telecommunications carrier shall be treated as a common carrier under this Act only to the extent that it is engaged in providing telecommunications services, except that the Commission shall determine whether the provision of fixed and mobile satellite service shall be treated as common carriage. [47 USC 153]

telecommunications common carrier

An individual, partnership, association, joint-stock company, trust or corporation engaged in rendering radio telecommunications services to the general public for hire. [47 CFR 22.99]

telecommunications device

(A) Shall not impose new obligations on broadcasting station licensees and cable operators covered by obscenity and indecency provisions elsewhere in this Act; and (B) does not include an interactive computer service. [47 USC 223]

telecommunications equipment

Equipment, other than customer premises equipment, used by a carrier to provide telecommunications services, and includes software integral to such equipment (including upgrades). [47 USC 153]

telecommunications industry

Communications businesses using regulated or unregulated facilities or services and includes broadcasting, telecommunications, cable, computer, data transmission, software, programming, advanced messaging, and electronics businesses. [47 USC 614]

telecommunications relay services [1]

Telephone transmission services that provide the ability for an individual who has a hearing impairment or speech impairment to engage in communication by wire or radio with a hearing individual in a manner that is functionally equivalent to the ability of an individual who does not have a hearing impairment or speech impairment to engage in communication by wire or radio with a hearing individual in a manner that is functionally equivalent to the ability of an individual who does not have a hearing impairment or speech impairment to communicate using voice communication services by wire or radio. Such term includes services that enable two-way communication between an individual who uses a TDD or other nonvoice terminal device and an individual who uses TDD or other nonvoice terminal and an individual who does not use such a device. [47 USC 225]

telecommunications relay services (TRS) [2]

Telephone transmission services that provide the ability for an individual who has a hearing or speech disability to engage in communication by wire or radio with a hearing individual in a manner that is functionally equivalent to the ability of an individual in a manner that is functionally equivalent to the ability of an individual who does not have a hearing or speech disability to communicate using voice communication services by wire or radio.   Such term includes services that enable two-way communication between an individual who uses a text telephone or other nonvoice terminal device and an individual who does not use such a device. TRS supersedes the terms "dual party relay system," "message relay services," and "TDD Relay." [47 CFR 64.601]

telecommunications service

The offering of telecommunications for a fee directly to the public, or to such classes of users as to be effectively available directly to the public, regardless of the facilities used. [47 USC 153]

telecommunications service priority (TSP) system user [1]

Any individual, organization, or activity that interacts with the NSEP TSP System. [47 CFR 64.1604]

telecommunications service priority (TSP) system user [2]

Any individual, organization, or activity that interacts with the TSP System. [47 CFR Pt. 216, App.]

telecommunications service provider

Interexchange carriers, operator service providers, enhanced service providers, and any other provider of interstate telecommunications services. [47 CFR 64.1201]

telecommunication services

The transmission, emission, or reception of signals, signs, writing, images, sounds, or intelligence of any nature, by wire, cable, satellite, fiber optics, laser, radio, visual, or other electronic, electric, eletromagnetic, or acoustically coupled means, or any combination thereof. The term can include necessary telecommunication facilities. [47 CFR Pt. 216, App., 47 CFR 64.1604]

telegram

Written matter intended to be transmitted by telegraphy for delivery to the addressee. This term also includes radiotelegrams unless otherwise specified. Note: In this definition the term telegraphy has the same general meaning as defined in the Convention. (CONV) [47 CFR 2.1]

telegraphy [1]

A form of telecommunication which is concerned in any process providing transmission and reproduction at a distance of documentary matter, such as written or printed matter or fixed images, or the reproduction at a distance of any kind of information in such a form. For the purposes of the [International] Radio Regulations, unless otherwise specified therein, telegraphy shall mean a form of telecommunication for the transmission of written matter by the use of a signal code. (RR) [47 CFR 2.1]

telegraphy [2]

A system of telecommunication for the transmission of written matter by the use of signal code. [47 CFR 21.2]

telemessaging service

Voice mail and voice storage and retrieval services, any live operator services used to record, transcribe, or relay messages (other than telecommunications relay services), and any ancillary services offered in combination with these services. [47 USC 260]

telemetering [1]

Automatic radiocommunication in a fixed or mobile service intended to indicate or record a measurable variable quantity at a distance. [47 CFR 21.2]

telemetering (also telemetry) [2]

The transmission of nonvoice signals for the purpose of automatically indicating or recording measurements at a distance from the measuring instrument. [47 CFR 90.7]

telemetry [1]

The use of telecommunication for automatical [sic] indicating or recording measurements at a distance from the measuring instrument. (RR) [47 CFR 2.1]

telemetry [2]

A one-way transmission of measurements at a distance from the measuring instrument. [47 CFR 97.3]

telephone

A terminal instrument which permits two-way, real-time voice communication with a distant party over a network or customer premises connection. It converts real-time voice and voiceband acoustic signals into electrical signals suitable for transmission over the telephone network and converts received electrical signals into acoustic signals. A telephone which meets the requirements of this standard also generates a magnetic field to which hearing aids may couple. [47 CFR 68.316]

telephone boundaries

The electrical interface with the network, PBX or KTS and the acoustic, magnetic and mechanical interfaces with the user. The telephone may also have an electrical interface with commercial power. [47 CFR 68.316]

telephone call

In §64.1200(a)(2) of this section shall not include a call or message by, or on behalf of, a caller (1) that is not made for a commercial purpose, (2) that is made for a commercial purpose but does not include the transmission of any unsolicited advertisement, (3) to any person with whom the caller has an established business relationship at the time the call is made, or (4) which is a tax-exempt nonprofit organization. [47 CFR 64.803]

telephone company

A carrier that provides telephone exchange service as defined in section 3(r) of the Communications Act of 1934. [47 CFR 69.2]

telephone connection

Connection to telephone network tip and ring leads for 2-wire and 4-wire connections and, additionally, for 4-wire telephone connecting, tip 1 and ring 1 leads and all connections derived from these leads. The term "derived" as used here means that the connections are not separated from telephone tip and ring or from telephone tip 1 and ring 1 by a sufficiently protective barrier. Part 68 Rules that apply specifically to telephone network tip and ring pairs shall also apply to telephone network tip 1 and ring 1 pais unless otherwise specified. In 4-wire connecting, leads designated tip and ring at the interface are for transmitting voice frequencies toward the network and leads designated tip 1 and ring 1 at the interface are for receiving voice frequencies from the network. [47 CFR 68.3]

telephone exchange service

(A) Service within a telephone exchange, or within a connected system of telephone exchanges within the same exchange area operated to furnish to subscribers intercommunicating service of the character ordinarily furnished by a single exchange, and which is covered by the exchange service charge, or (B) comparable service provided through a system of switches, transmission equipment, or other facilities (or combination thereof) by which a subscriber can originate and terminate a telecommunications service. [47 USC 153]

telephone facsimile machine

Equipment which has the capacity (A) to transcribe text or images, or both, from paper into an electronic signal and to transmit that signal over a regular text or images, or both, from paper into an electronic signal and to transmit that signal over a regular telephone line, or (B) to transcribe text or images (or both) from an electronic signal received over a regular telephone line onto paper. [47 USC 227]

telephone network

The public switched network and those private lines which are defined in §68.2(a) (2) and (3). [47 CFR 68.3]

telephone service area [1]

When used in connection with a common carrier subject in whole or in part to title II of this Act, the area within which such carrier provided telephone exchange service as of January 1, 1993, but if any common carrier after such date transfers its telephone exchange service facilities to another common carrier, the area to which such facilities provide telephone exchange service shall be treated as part of the telephone service area of the acquiring common carrier and not of the selling common carrier. [47 USC 572]

telephone service area [2]

When used in connection with a common carrier subject in whole or in part to title II of this Act means the area within which such carrier is offering telephone exchange service. [47 USC 653]

telephone solicitation

The initiation of a telephone call or message for the purpose of encouraging the purchase or rental of, or investment in, property, goods, or services, which is transmitted to any person, but such term does not include a call or message (A) to any person with that person's prior express invitation or permission, (B) to any person with whom the caller has an established business relationship, or (C) by a tax exempt nonprofit organization. [47 USC 227, 64.1200]

telephone toll service

Telephone service between stations in different exchange areas for which there is made a separate charge not included in contracts with subscribers for exchange service. [47 USC 153]

telephony

A form of telecommunication set up for the transmission of speech or, in some cases, other sounds. (RR) [47 CFR 2.1, 21.2]

television

A form of telecommunication for the transmission of transient images of fixed or moving objects. (RR) [47 CFR 2.1, 21.2]

television broadcast band

The frequencies in the band extending from 54 to 806 megahertz which are assignable to television broadcast stations. These frequencies are 54 to 72 megahertz (channels 2 through 4), 76 to 88 megahertz (channels 5 and 6), 174 to 216 megahertz (channels 7 through 13), and 470 to 806 megahertz (channels 14 through 69). [47 CFR 73.681]

television broadcast booster station

A station in the broadcast service operated by the licensee or permittee of a full service television broadcast station for the purpose of retransmitting the programs and signals of such primary station without significantly altering any characteristic of the original signal other than its amplitude. A television broadcast booster station may only be located such that its entire service area is located within the protected contour of the primary station it retransmits. For purposes of this paragraph, the service area of the booster and the protected contour of the primary station will be determined by the methods prescribed in §74.705 (c). [47 CFR 74.701]

television (TV) broadcast receiver

A device designed to receive television pictures that are broadcast simultaneously with sound on the television channels unauthorized under part 73 of this chapter. [47 CFR 15.3]

television broadcast station

A station in the television broadcast bend transmitting simultaneous visual and aural signals intended to be received by the general public. [47 73.681]

television broadcast translator station

A station in the broadcast service operated for the purpose of retransmitting the programs and signals of a television broadcast station, without significantly altering any characteristic of the original signal other than its frequency and amplitude, for the purpose of providing television reception to the general public. [47 CFR 74.701]

television channel

A band of frequencies 6 MHZ wide in the television broadcast band and designated either by number or by the extreme lower and upper frequencies. [47 CFR 73.681]

television market

For purposes of the must-carry rules

(1) A local commercial broadcast television station's market shall be defined as its Area of Dominant Influence (ADI) as determined by Arbitron and published in its Television ADI Market Guide or any successor publication, as noted below, except that for areas outside the contiguous 48 states the area of dominant influence may be defined using Nielsen's Designated Market Area (DMA), where applicable, and that Puerto Rico, the U.S. Virgin Islands and Guam will each be considered one ADI;

(2) A cable system's television market(s) shall be the one or more ADIs in which the communities it serves are located;

(3) In addition, the county in which a station's community of license is located will be considered within its market. Note: For the 1993 must-carry/retransmission consent election, the ADI assignments specified in the 1991-1992 *Television Market Guide* will apply. [47 CFR 76.55]

television network

Any person, entity, or corporation providing on a regular basis more than fifteen (15) hours of prime time programming per week (exclusive of live coverage of bona fide news events of national importance) to interconnected affiliates that reach, in aggregate, at least seventy-five (75) percent of television households nationwide; and/or any person, entity, or corporation controlling, controlled by, or under common control with such person, entity, or corporation. Not included within this definition is any television network formed for the purpose of producing, distributing, or syndicating program material for educational, noncommercial, or public broadcasting exhibition, or for non-English language exhibition, or that predominately distributes programming involving the direct sale of products or services. [47 CFR 73.662]

television nonbroadcast pickup station

A mobile, except television pickup, station used for the transmission of television program material and related communications for non-broadcast purposes. [47 CFR 21.2]

television pickup station

A land mobile station used for the transmission of television program material and related communications from the scenes of events occurring at points removed from television broadcast station studios to television broadcast stations. [47 CFR 21.2]

television program producer

Television program producer refers to a person or organization engaged in the production of television programs. [47 CFR 74.801]

television station; television broadcast station

Any television broadcast station operating on a channel regularly assigned to its community by §73.606 of this chapter, and any television broadcast station licensed by a foreign government; provided, however, That a television broadcast station licensed by foreign government shall not be entitled to assert a claim to carriage, program exclusivity, or retransmission consent authorization pursuant to subpart D or F of this part, but may otherwise be carried if consistent with the rules on any service tier. [47 CFR 76.5]

television STL station (studio transmitter link)

A fixed station used for the transmission of television program material and related communications from a studio to the transmitter of a television broadcast station. [47 CFR 21.2]

television translator station

A television broadcast translator station as defined in §74.701 of this chapter. [47 CFR 76.5]

television transmission standards

The standards which determine the characteristics of a television signal as radiated by a television broadcast station. [47 CFR 73.681]

television transmitter

The radio transmitter or transmitters for the transmission of both visual and aural signals.   [47 CFR 73.681]

temporary fixed ITFS station

An ITFS station used for the transmission of material from temporary unspecified points to an ITFS station. [47 CFR 74.901]

temporary fixed station
> One or more fixed transmitters that normally do not remain at any particular location for longer than 6 months. [47 CFR 22.99]

temporary service
> Service for a period not exceeding 6 months. [47 CFR 63.04]

temporary unspecified
> Whenever a station is to transmit from unspecified locations within a prescribed geographical area, the station location is *temporary unspecified* and the proposed geographical operating area must be shown on the application. [47 CFR 80.39]

terminal isolation
> The attenuation, at any subscriber terminal, between that terminal and any other subscriber terminal in the cable television system. [47 CFR 76.5]

terminal port
> An equipment port of registered protective circuitry which port faces remotely located terminal equipment. [47 CFR 68.3]

terrestrial radiocommunication
> Any radiocommunication other than space radiocommunication or radio astronomy. (RR) [47 CFR 2.1, 25.201]

terrestrial station
> A station effecting terrestrial radiocommunication. Note: In these [International Radio] Regulations, unless otherwise stated, any station is a terrestrial station. (RR) [47 CFR 2.1, 25.201]

test equipment
> Equipment connected at the customer's premises that is used on the customer's side of the network interfaces: (a) to measure characteristics of the telephone network; or (b) to detect and/or isolate a communications fault between a terminal equipment entity and the telephone network. Registration is required for test equipment capable of functioning as temporary terminal equipment, such as portable traffic recorders or equipment capable of transmitting test tones, either as generators or responders. [47 CFR 68.3]

text

When written with an upper-case "T", refers to the Text Mode. When written with a lower-case "t", refers to any combination of displayable characters. [47 CFR 15.119]

text change

A change in the text of a tariff which does not result in a change in any rate or regulation. [47 CFR 61.3]

text telephone (ITT)

A machine that employs graphic communication in the transmission of coded signals through a wire or radio communication system. TT supersedes the term "TDD" or "telecommunications device for the deaf." [47 CFR 64.601]

thermal noise power

The noise power in watts defined by the formula N=kTB where N is the noise power in watts, K is Boltzmann's constant, T is the absolute temperature in degrees Kelvin (*e.g.*, 295°K) and B is the emission bandwidth of the device in hertz. [47 CFR 15.303]

third party communications

A message from the control operator (first party) of an amateur station to another amateur station control operator (second party) on behalf of another person (third party). [47 CFR 97.3]

TIA

Telecommunications Industry Association [47 CFR 22.933]

tie trunk transmission interfaces

(a) 2-wire: a 2-wire transmission interface with a path that is essentially lossless (except for 2dB switched pad operation, or equivalent) between the interface and the 2-wire or 4-wire, transmission reference point of the terminal equipment.

(b) 4-wire lossless: A 4-wire transmission interface with a path that is essentially lossless (except for 2dB switched pad operation, or equivalent) between the interface and the 2-wire or 4-wire transmission reference point of the terminal equipment; and

(c) 4-wire conventional terminating set (CTS): A 4-wire interface with a path to the transmission reference point that has a conventional terminating set providing 2-wire to 4-wire conversion with approximately 4dB of loss and having no gain elements. This device's loss will be referred to as a "nominal" 4dB, but in no case is it allowed to be less than 3dB.

(d) Direct digital interface: An interface between a digital PBX and a digital transmission facility.

(e) Digital tandem 4-wire interface: A 4-wire digital interface between digital terminal equipment and a digital transmission facility operating at 1.544 Mbps or subrate connecting terminal equipment that provide tandem connections.

(f) Digital satellite 4-wire interface: A 4-wire digital interface between digital terminal equipment and a digital transmission facility operating at 1.544 Mbps or subrate connecting terminal equipment that does not provide tandem connections to other digital terminal equipment. [47 CFR 68.3]

time brokerage

The sale by a licensee of discrete block of time to a "broker" that supplies the programming to fill that time and sells the commercial spot announcements in it. [47 CFR 73.3555]

time hopping systems

A time hopping system is a spread spectrum system in which the period and duty cycle of a pulsed RF carrier are varied in a pseudorandom manner under the control of a coded sequence. Time hopping is often used effectively with frequency hopping to form a hybrid time-division, multiple-access (TDMA) spread spectrum system. [47 CFR 2.1]

time of installation

The date at which telecommunications plant is placed in service. [47 CFR 32.9000]

time of retirement

The date at which telecommunications plant is retired from service. [47 CFR 32.9000]

time window

An interval of time in which transmission is desired. [47 CFR 15.303]

TOA

Tornado watch [47 CFR 11.31]

toll center

An office (or group of offices) within a city which generally handles the originating and incoming toll traffic for that city to or from other toll center areas and which handles through switched traffic. The toll center normally handles the inward toll traffic for its tributary exchanges and, in general, either handles the outward traffic originating at its tributaries or serves as the outlet to interexchange circuits for outward traffic ticketed and timed at its tributaries. Toll centers are listed as such in the Toll Rate and Route Guide. [47 CFR Pt. 36, App.]

toll center area

The areas served by a toll center, including the toll center city and the communities served by tributaries of the toll center. [47 CFR Pt. 36, App.]

toll center toll office

A toll office (as contrasted to a local office) in a toll center city. [47 CFR Pt. 36, App.]

toll circuit

A general term applied to interexchange trunks used primarily for toll traffic. [47 CFR Pt. 36, App.]

toll connecting trunk

A general classification of trunks carrying toll traffic and ordinarily extending between a local office and a toll office, except trunks classified as tributary circuits. Examples of toll connecting trunks include toll switching trunks, recording trunks and recording-completing trunks. [47 CFR Pt. 36, App.]

toll office

A central office used primarily for supervising and switching toll traffic. [47 CFR Pt. 36, App.]

TOR

Tornado warning [47 CFR 11.31]

total assets
The book value (except where generally accepted accounting principles (GAAP) require market valuations of all property owned by an entity, whether real or personal, tangible or intangible, as evidenced by the most recent audited financial statements. [47 CFR 26.4, 24.720]

TP
RF transmitter power expressed in W, either mean or peak envelope, as measured at the transmitter output antenna terminals. [47 CFR 95.669]

TPO
Transmitter power output [47 CFR 74.1235]

T/R
Connections to the "tip" and "ring" wires of a telephone communications line, trunk, channel or facility. [47 CFR 68.502]

traffic over first routes
A term applied to the routing of traffic and denoting routing via principal route for traffic between any two points as distinguished from alternate routes for such traffic. [47 CFR Pt. 36, App.]

transfer of control
A transfer of the controlling interest in a Public Mobile Services licensee from one party to another. [47 CFR 22.99]

transfer switch
A device used to alternate between the reception of over-the -air radio frequency signals via connection to an antenna and the reception of radio frequency signals received by any other method, such as from a TV interface device. [47 CFR 15.3]

transitional support (TRS)
Funds provided by telephone companies that are not association Common Line tariff participants, but were net contributors to the association Common Line pool in 1988, to telephone companies that are not association Common Line tariff participants and were net receivers from the association Common Line pool in 1988. [47 CFR 69.2]

translator coverage contour

The coverage contour for an FM translator providing "fill-in" service is congruent with its parent station: For a fill-in translator for a commercial Class B station it is the predicted 0.5 mV/m field strength contour; for a fill-in translator for a commercial Cass B1 station it is the predicted 0.7 mV/m field strength contour; and for a fill-in translator for all other classes of commercial stations as well as all noncommercial educational stations it is the predicted 1 mV/m field strength contour. A fill-in FM translator's coverage contour must be contained within the primary station's coverage contour. The protected contour for an FM translator station is its predicted 1 mV/m contour. [47 CFR 74.1201]

transmission of energy by radio (radio transmission of energy)

Includes both such transmission and all instrumentalities, facilities, and services incidental to such transmission. [47 USC 153]

transmitter [1]

The transmitter unit and all auxiliary equipment necessary to make this unit operate as a main or emergency transmitter in a ship station at sea. Each separate motor-generator, rectifier, or other unit required to convert the ship primary power to the phase, frequency, or voltage necessary to energize the transmitter unit is considered a component of the transmitter. [47 CFR 80.251]

transmitter [2]

Apparatus that converts electrical energy received from a source into RF energy capable of being radiated. [47 CFR 95.669]

transmitter-hour

One frequency used on one transmitter for one hour. [47 CFR 73.701]

transparent space

Transmitted as a special character, it is a one-column-wide space behind which program video is always visible (except when a transparent space immediately precedes or follows a displayable character and solid box is needed to make that character legible. [47 CFR 15.119]

transponder

A transmitter-receiver facility the function of which is to transmit signals automatically when the proper interrogation is received. (FCC) [47 CFR 2.1]

travelers' information station

A base station in the Local Government Radio Service used to transmit non-commercial, voice information pertaining to traffic and road conditions, traffic hazard and traveler advisories, directions, availability of lodging, rest stops, and service stations, and descriptions of local points of interest. [47 CFR 90.7]

tributary circuit

A circuit between a tributary office and a toll switchboard or intertoll dialing equipment in a toll center city. [47 CFR Pt. 36, App.]

tributary office

A local office which is located outside the exchange in which a toll center is located, which has a different rate center from its toll center and which usually tickets and times only a part of its originating toll traffic, but which may ticket or time all or none of such traffic. The toll center handles all outward traffic not ticketed and timed as the tributary and normally switches all inward toll traffic from outside the tributary's toll center to the tributary. Tributary offices are indicated as such in the Toll Rate and Route Guide. [47 CFR Pt. 36, App.]

tropospheric scatter

The propagation of radio waves by scattering as a result of irregularities or discontinuities in the physical properties of the troposphere. (RR) [47 CFR 2.1]

TRS

Transitional support [47 CFR 69.2]

TRS

Telecommunications relay services [47 CFR 64.601]

trunk (see line)

trunk (telephony)

A one or two-way channel provided as a common traffic artery between switching equipment. [47 CFR 90.7]

trunk group

All of the trunks of a given type of characteristic that extend between two switching points. [47 CFR 90.7]

trunked radio system

A method of operation in which a number of radio frequency channel pairs are assigned to mobile and base stations in the system for use as a trunk group. [47 CFR 90.7]

trunks

Circuit between switchboards or other switching equipment, as distinguished from circuits which extend between central office switching equipment and information origination/termination equipment. [47 CFR Pt. 36, App.]

TSA

Tsunami watch [47 CFR 11.31]

TSO

Technical standard order [47 CFR 87.5]

TSP

Telecommunications service priority [47 CFR 64.1604]

TSPS

Traffic service position system [47 CFR 36.123]

TSPS complex

All groups of operator positions, wherever located, associated with the same TSPS stored program control units. [47 CFR Pt. 36, App.]

TSW

Tsunami warning [47 CFR 11.31]

TT

Text telephone [47 CFR 64.601]

TV

Television [47 CFR 15.3]

TV broadcast licensee

Licensees and permittees of both TV broadcast and low power TV stations, unless specifically otherwise indicated. [47 CFR 74.601]

TV broadcast station or TV station

Excludes stations which are primarily satellite operations. [47 CFR 73.3555]

TV interface device

An unintentional radiator that produces or translates in frequency a radio frequency carrier modulated by a video signal derived from an external or internal signal source, and which feeds the modulated radio frequency energy by conduction to the antenna terminals or other non-baseband input connections of a television broadcast receiver. A TV interface device may include a standalone RF modulator, or a composite device consisting of an RF modulator, video source and other components devices. Examples of TV interface devices are video cassette recorders and terminal devices attached to a cable system or used with a Master Antenna (including those used for central distribution video devices in apartment or office buildings). [47 CFR 15.3]

TV microwave booster station

A fixed station in the TV broadcast auxiliary service that receives and amplifies signals of a TV pickup, TV STL, TV relay, or TV translator relay station and retransmits them on the same frequency. [47 CFR 74.601]

TV pickup stations

A land mobile station used for the transmission of TV program material and related communications from scenes of events occurring at points removed from TV broadcast station studios to TV broadcast or low power TV stations or other purposes as authorized in §74.631. [47 CFR 74.601]

TV relay station

A fixed station used for transmission of TV program material and related communications for use by TV broadcast and lower power TV stations or other purposes as authorized in §74.631. [47 CFR 74.601]

TV STL station (studio-transmitter link)

A fixed station used for the transmission of TV program material and related communications from the studio to the transmitter of a TV broadcast or low power TV station or other purposes as authorized in §74.631. [47 CFR 74.601]

TV translator relay station

A fixed station used for relaying programs and signals of TV broadcast stations to LPTV, TV translator, and to other communications facilities that the Commission may authorize or for other purposes as permitted by §74.631. [47 CFR 74.601]

TWX

Teletypewriter exchange service. [47 CFR Pt. 36, App.]

TWX connection

A completed TWX call, *i.e.*, a call on which a TWX communication was passed between the calling and called stations. [47 CFR Pt. 36, App.]

TWX connection-minute-kilometers

The product of (a) the number of TWX connections, (b) the average minutes per TWX connection and (c) the average route kilometers of circuits involved. [47 CFR Pt. 36, App.]

TWX switching plant trunks

Interexchange circuits, excluding remote access lines, which handle 100 word per minute TWX traffic only. [47 CFR Pt. 36, App.]

TX

Transmitter [47 CFR 2.1003]

U

UHF

Ultra-high frequency [47 CFR 97.3]

UHF translator

A television broadcast translator station operating on a UHF television broadcast channel. [47 CFR 74.701]

UHF translator signal booster

A station in the broadcasting service operated for the sole purpose of retransmitting the signals of the UHF translator station by amplifying and reradiating such signals which have been received directly through space, without significantly altering any characteristic of the incoming signal other than its amplitude. [47 CFR 74.701]

ultra-high frequency (UHF)

The frequency range 300-3000 MHZ. [47 CFR 97.3]

ultrasonic equipment

A category of ISM equipment in which the RF energy is used to excite or drive an electromechanical transducer for the production of sonic or ultrasonic mechanical energy for industrial, scientific, medical or other noncommunication purposes. [47 CFR 18.107]

unattended operation

Operation of a station by automatic means whereby the transmitter is turned on and off and performs its functions without attention by a qualified operator. [47 CFR 74.901, 78.5]

underway

A vessel is underway when it is not at anchor, made fast to the shore, or aground. [47 CFR 80.5]

undue burden

Significant difficulty or expense. In determining whether the closed captions necessary to comply with the requirements of this paragraph would result in an undue economic burden, the factors to be considered include (1) the nature and cost of the closed captions for the programming; (2) the impact on the operation of the provider or program owner; (3) the financial resources of the provider or program owner; and (4) the type of operations of the provider or program owner. [47 USC 613]

uniformed services

Meaning given that term by section 101(3) of title 37 of the United State Code. [47 CFR 19.735-102]

unintentional radiator

A device that intentionally generates radio frequency energy for use within the device, or that sends radio frequency signals by conduction to associated equipment via connecting wiring, but which is not intended to emit RF energy by radiation or induction. [47 CFR 15.3]

unit of capacity

The capability to transmit one conversation. [47 CFR 69.2]

United States [1]

The several states and territories, the District of Columbia, and the possessions of the United States, but does not include [the Philippine Islands or] the Canal Zone. [47 USC 153]

United States [2]

The several states and territories, the District of Columbia, and the possessions of the United States. [47 CFR 61.3]

unlicensed wireless service
The offering of telecommunications services using duly authorized devices which do not require individual licenses, but does not mean the provision of direct-to-home satellite services (as defined in Section 303(v). [47 USC 332]

unprotected non-system premises wiring
All other non-system premises wiring. [47 CFR 68.3]

unprotected system premises wiring
All other premises wiring. [47 CFR 68.3]

unsecured credit
The furnishing of service without maintaining on a continuing basis advance payment, deposit, or other security, that is designed to assure payment of the estimated amount of service for each future 2 months period, with revised estimates to be made on at least a monthly basis. [47 CFR 64.803]

unserved areas
In the Cellular Radiotelephone Service, areas outside of all existing CGSAs (on either of the channel blocks), to which the Communications Act of 1934, as amended, is applicable. [47 CFR 22.99]

unserved household
Has the meaning given those terms, respectively, in Section 119(d) of title 17, United States Code, as in effect on the date of enactment of the Cable Television Consumer Protection and Competition Act of 1992. [47 USC 325]

unsolicited advertisement
Any material advertising the commercial availability or quality of any property, goods, or services which is transmitted to any person without that person's prior express invitation or permission. [47 USC 227, 47 CFR 64.1200]

unwanted emissions
Consist of spurious emissions and out-of-band emissions. (RR) [47 CFR 2.1]

UPI

United Press International [47 CFR 11.43]

urbanized area

A city and the surrounding closely settled territories. [47 CFR 90.7]

urgency signal

(1) The urgency signal is the international radiotelegraph or radiotelephone signal which indicates that the calling station has a very urgent message to transmit concerning the safety of a ship, aircraft, or other vehicle, or of some person on board or within sight.

(2) In radiotelegraphy, the international urgency signal consists of three repetitions of the group "XXX", sent before the call, with the letters of each group and the successive groups clearly separated from each other.

(3) In radiotelephony, the international urgency signal consists of three oral repetitions of the groups of words "PAN PAN", each word of the group pronounced as the French word "PANNE" and sent before the call. [47 CFR 80.5]

usable activated channels

Activated channels of a cable system, except those channels whose use for the distribution of broadcast signals would conflict with technical and safety regulations as determined by the Commission. [47 USC 522, 47 CFR 76.5]

usable space

The space on a utility pole above the minimum grade level which can be used for the attachment of wires, cables, and associated equipment. [47 CFR 1.1402]

USC

United States Code [47 CFR 0.457]

uses

As used in this section and 73.1942, a candidate appearance (including by voice or picture) that is not exempt under paragraphs 73.1941 (a)(1) through (a)(4) of this section. [47 CFR 73.1941]

USGS

United States Geological Survey [47 CFR 24.53]

USIA

United States Information Agency [47 CFR 1.1701]

USOA

Uniform system of accounts [47 CFR 32.1]

USOC

Universal service ordering code [47 CFR 68.502]

UTC

Coordinated universal time [47 CFR 73.702]

utility

Any person whose rates or charges are regulated by the federal government or a state and who owns or controls poles, ducts, conduits, or rights-of-way used, in whole or in part, for wire communications. Such term does not include any railroad, any person who is cooperatively organized, or any person owned by the federal government or any state. [47 CFR 1.1402]

V

VBI

Vertical blanking intermission [47 CFR 73.646]

VCO

Voice carry over [47 CFR 64.601]

VE

Volunteer examiners [47 CFR 1.912]

VEC

Volunteer-examiner coordinator [47 CFR 1.934]

very high frequency (VHF)

The frequency range of 30-300 MHZ. [47 CFR 97.3]

vessel (see ship)

vessel traffic service (VTS)

A U.S. Coast Guard traffic control service for ships in designated water areas to prevent collisions, groundings and environmental harm. [47 CFR 80.5]

vestigial sideband transmission

A system of transmission wherein one of the generated sidebands is partially attenuated at the transmitter and radiated only in part. [47 CFR 73.681]

VHF

Very high frequency [47 CFR 97.3]

VHF omni directional range station (VOR)
A radionavigation land station in the aeronautical radionavigation service providing direct indication of the bearing (omni-bearing) of that station from an aircraft. [47 CFR 87.5]

VHF translator
A television broadcast translator station operating on a VHF television broadcast channel. [47 CFR 74.701]

VHF/UHF
Very high frequency/ultra high frequency [47 CFR 15.117]

video description
The insertion of audio narrated descriptions of a television program's key visual elements into natural pauses between the program's dialogue. [47 USC 613]

video entertainment material
The transmission of a video signal (*e.g.* United States Standard Monochrome or National Television Systems Committee 525-line television) and an associated audio signal which is designed primarily to amuse or entertain, such as movies and games. [47 CFR 94.3]

video programming
Programming provided by, or generally considered comparable to programming provided by, a television broadcast station. [47 USC 522]

video programming vendor
A person engaged in the production, creating, or wholesale distribution of video programming for sale. [47 CFR 76.1300]

visual carrier frequency
The frequency of the carrier which is modulated by the picture information. [47 CFR 73.681]

visual signal level
The rms voltage produced by the visual signal during the transmission of synchronizing pulses. [47 CFR 76.5]

visual transmissions

Communications or message transmitted on a subcarrier intended for reception and visual presentation on a viewing screen, teleprinter, facsimile printer, or other form of graphic display or record. [47 CFR 73.310]

visual transmitter

The radio equipment for the transmission of the visual signal only. [47 CFR 73.681]

visual transmitter power

The peak power output when transmitting a standard television signal. [47 CFR 73.681]

VOA

Voice of America [47 CFR 73.1207]

voiceband metallic private line channel interface

The point of connection between a voiceband metallic private line channel and terminal equipment or systems where the network does not provide any signaling or transmission enhancement. Registered terminal equipment or systems may use convenient signaling methods so long as the signals are provided in such a manner that they cannot interfere with adjacent network channels. All tip and ring leads shall be treated as telephone connecting for the purpose of fulfilling registration conditions. [47 CFR 68.3]

voice carry over (VCO)

A reduced form of TRS where the person with the hearing disability is able to speak directly to the other end user. The CA types the response back to the person with the hearing disability. The CA does not voice the conversation. [47 CFR 64.601

VOR

VHF omni direction range station [47 CFR 87.5]

VSAT

Very small aperture terminal [47 CFR 25.134]

VTS

Vessel traffic service [47 CFR 80.5]

W

W
  watts [47 CFR 95.669]

waiver
  The cancellation, remission, forgiveness, or non-recovery of a debt allegedly owed by an employee to an agency as permitted or required by 5 U.S.C. 5584, 10 U.S.C. 2774, or 32 U.S.C. 710, 5 U.S.C. 8346(b), or any other law. [47 CFR 1.1901]

WASP
  War Air Service Program [47 CFR 87.393]

watch
  The act of listening on a designated frequency. [47 CFR 80.5]

WATS
  Wide area telephone service [47 CFR 64.1604]

WATS Access Line
  A line or trunk that is used exclusively for WATS service. [47 CFR 69.2]

weighted standard work second
  A measurement of traffic operating work which is used to express the relative time required to handle the various kinds of calls or work functions, and which is weighted to reflect appropriate degrees of waiting to serve time. [47 CFR Pt. 36, App.]

white area

The area or population which does not receive interference-free primary service from an authorized AM station or does not receive a signal strength of at least 1 mV/m from an authorized FM station. [47 CFR 73.14]

wide area telephone service WATS

A toll service offering for customer dial type telecommunications between a given customer station and stations within specified geographic rate areas employing a single access line between the customer location and the serving central office. Each access line may be arranged for either outward (OUT-WATS) or inward (IN-WATS) service or both. [47 CFR Pt. 36, App.]

wideband channel

A communication channel of a bandwidth equivalent to twelve or more voice grade channels. [47 CFR Pt. 36, App.]

willful

With reference to the commission or omission of any act, the conscious and deliberate commission or omission of such act, irrespective of any intent to violate any provision of this Act or any rule or regulation of the Commission authorized by this Act or by a treaty ratified by the United States. [47 USC 312]

wire communication (communication by wire)

The transmission of writing, signs, signals, pictures, and sounds of all kinds by aid of wire, cable, or other like connection between the points of origin and reception of such transmission, including all instrumentalities, facilities, apparatus and services (among other things, the receipt, forwarding, and delivery of communications) incidental to such transmission. [47 USC 153]

wireline common carrier

A telecommunications common carrier that is also engaged in the business of providing landline local exchange telephone service. [47 CFR 22.99]

working loop

A revenue-producing pair of wire, or its equivalent, between a customer's station and the central office from which the station is served. [47 CFR Pt. 36, App.]

WRSAME

    NWS weather radio specific area message encoder [47 CFR 11.31]

WSA

    Winter storm watch [47 CFR 11.31]

WSW

    Winter storm warning [47 CFR 11.31]

WXR

    National Weather Service [47 CFR 11.31]

X Y Z

XMTR
    Transmitter [47 CFR 2.1003]

zero level decoder
    A decoder that yields an analog level of 0 dBm at its output when the input is the digital milliwatt signal. See Figure 68.3(j). [47 CFR 68.3]

zone of reception
    Any geographic zone indicated in §73.703 in which the reception of particular programs in specifically intended and in which broadcast coverage is contemplated. [47 CFR 73.701]

Telecommunications Act Handbook

From the Foreword...

"[The Telecommunications Act Handbook] is not simply a dry recitation of the provisions of the new telecommunications law. Instead, it puts the new law in context—helping telecommunications professionals and laymen alike understand why it's important, and how it will affect their lives."

Congressman Jack Fields,
*Chairman, Subcommittee on
Telecommunications and Finance
U.S. House of Representatives*

The Telecommunications Act of 1996 will dramatically change the way you do business in the telecommunications industry. Its impact on the market will result in huge transferals of wealth.

Is your company prepared to claim its market share?

The new **Telecommunications Act Handbook:** *A Complete Reference for Business* provides practical, non-legalese explanations of this legal framework, and delves into the challenges, risks, and benefits confronting this market. This comprehensive handbook also summarizes the history and development of each sector of the telecommunications industry, and then clearly explains how the new legislation will change each industry sector.

Don't allow your company to miss out on this $1 trillion dollar opportunity. With the **Telecommunications Act Handbook:** *A Complete Reference to Business,* you'll quickly learn the enormous impact of the law that will govern telecommunication activities for years to come.

Telecommunications Act Handbook: *A Complete Reference for Business*
Hardcover, Index, 640 pages, 1996, ISBN: 0-86587-545-6 **$89**

Government Institutes

4 Research Place, Rockville Maryland 20850
tel. (301) 921-2355 fax (301) 921-0373
Website- http://www.govinst.com
E-mail- giinfo@govinst.com

Contact us for a complete catalog of our books.

Government Institutes

4 Research Place • Rockville, Maryland 20850 •
tel. (301) 921-2355 • fax (301) 921-0373

Telecommunications: Glossary of Telecommunication Terms
By **National Telecommunications and Information Administration**

This new glossary contains over 5,000 technical terms and definitions that were standardized by the federal government as Federal Standard FED-STD-1037C. Telecommunication specialists in the U.S. and throughout the world have used this standard since the completion of the document in August 1996. It includes national and international terms draws from the International Telecommunications Union, the International Organization for Standardization, the TIA, ANSI, and others.

Softcover, approx. 400 pages, March 1997, ISBN: 0-86587-580-4 **$69**

Internet and the Law: *Legal Fundamentals for the Internet User*
By Raymond A. Kurz,**Rothwell, Figg, Ernst & Kurz**

The Explosion in the use of the Internet has thrust unsuspecting, but well-intentioned companies, organizations, and individuals (like you) into potential legal nightmares. But until now, there has been nos source of information to help you, the Internet user, avoid unnecessary conflicts.

This book provides you with an understanding of the legal landscape within which you operate. The basic principles pertaining to laws of copyright, trademark, trade secret, patent, libel/defamation and related issues as well as the basic principles of licensing are explained. This book outlines steps you can take to avoid or minimize your chances of unknowingly engaging in unlawful activity. In addition, this book helps assure that your rights as an Internet user will be protected.

Hardcover, 249 pages, 1996, ISBN: 0-86587-506-5 **$75**